D0844168

Psychotherapy and Buddhism

Toward an Integration

Issues in the Practice of Psychology

SERIES EDITOR:

George Stricker, *Derner Institute of Advanced Psychological Studies, Adelphi University, Garden City, New York*

PSYCHOTHERAPY AND BUDDHISM: Toward an Integration
Jeffrey B. Rubin

A Continuation Order Plan is available for this series. A continuation order will bring delivery of each new volume immediately upon publication. Volumes are billed only upon actual shipment. For further information please contact the publisher.

Psychotherapy and Buddhism

Toward an Integration

JEFFREY B. RUBIN

Private Practice in Psychoanalysis and Psychotherapy
New York, New York, and Bedford Hills, New York

PLENUM PRESS • NEW YORK AND LONDON

Library of Congress Cataloging-in-Publication Data

Rubin, Jeffrey B.
    Psychotherapy and Buddhism : toward an integration / Jeffrey B.
Rubin.
        p.   cm. -- (Issues in the practice of psychology)
    Includes bibliographical references and index.
    ISBN 0-306-45441-6
    1. Psychotherapy--Religious aspects--Buddhism. 2. Buddhism-
-Psychology.   I. Title. II. Series.
BQ4570.P76R83   1996ESS
294.3'375--dc20   96-34822
                                                                      CIP

ISBN 0-306-45441-6

© 1996 Plenum Press, New York
A Division of Plenum Publishing Corporation
233 Spring Street, New York, N. Y. 10013

10 9 8 7 6 5 4 3 2 1

Printed in the United States of America

To Joel and Mary, sine qua non

Preface

Psychoanalysis and Buddhism each have something rare and vital to contribute to the difficulties and challenges of living in our world. The capacity of these two wisdom traditions to help us live with greater self-awareness, self-acceptance, care, and freedom is essential in a world permeated by self-blindness, self-hatred, powerlessness, insensitivity, and alienation.

There has been a greater interest in the possible relationship between Western psychology and Eastern spiritual disciplines in recent years. In articles in professional and popular magazines, at conferences among Western mental health professionals and spiritual seekers and teachers, and on television specials on Eastern thought and healing there is much talk of a "dialogue" between Eastern and Western approaches to healing and self-transformation.

In the 18 years that I have been exploring psychoanalysis and Buddhism in tandem, I have repeatedly witnessed, within myself and others, their capacity to enlighten and enrich each other. And yet, the contact between Western psychology and Eastern spirituality usually resembles a monologue rather than a dialogue, in which one tradition is presumed to be superior to the other and enjoys a privileged status. Ascertaining the value of either tradition is not possible when either psychotherapy or Buddhism is viewed as superior to the other. The value of the devalued tradition is then neglected.

In this book I explore the relationship between psychotherapy and Buddhism from a very different perspective. Believing that each has something essential to contribute to the challenges of living, I view them as equal partners in a dialogue designed to explore their respective strengths and limitations. My hope is that this will generate a new sort of relationship and conversation that will illuminate a variety of topics that might be of interest to Buddhists and other spiritual seekers, interested laypeople, and mental health professionals and clients, such as the nature of health and illness, the path to self-awareness and self-transformation, the obstacles to change, and the possibilities of living with greater awareness, self-acceptance, tolerance, and inner peace. My ultimate hope is that this volume contributes—even in a small way—to increasing empathy for self and others, lessening human suffering and imprisonment, and fostering human liberation and freedom.

Acknowledgments

Scholarship is, for me, both a solitary and a communal enterprise. In the 20 years that I have been exploring psychoanalysis and Buddhism many people have provided intellectual, emotional, and spiritual support and guidance. I am deeply grateful to all those people from whom I have learned about psychoanalysis, Buddhism, and myself.

Jack Kornfield, Joseph Goldstein, and Christopher Titmuss have sensitively and skillfully guided me through the complexities and depths of Buddhist meditation. They have deeply enriched my understanding of Buddhist theory and practice and myself. I am deeply grateful for their generosity, patience, clarity, and wisdom. Their teachings and examples have been enormously important on a personal as well as professional level.

My psychoanalytic education has come from many quarters, including psychoanalytic patients, teachers, supervisors, and analysts, too numerous to cite specifically, too important not to mention. To them I am deeply grateful.

Of the many people who have taught me about psychoanalytic theory and practice, I would like to single out George Atwood. I have been inspired by his love of psychoanalysis, his appreciation of the complexities of human subjectivity, and his intellectual generosity. He has read large chunks of this manuscript and discussed most, if not all, of the ideas in this work with interest and discernment. His intellectual humility and nonattachment, combined with his critical acumen, were a living embodiment of the integration of Eastern and Western modes of thinking. They were also enormously empowering on a personal as well as professional level.

Zvi Lothane had a seminal role in the origins of this work when he read what is now Chapter 6 in the early 1980s and encouraged me to publish it. I am grateful for his intellectual generosity and personal support over the years.

I also owe a debt of gratitude to my patients, who have taught me about the intricacies and complexities of human nature, the therapeutic process, and myself. This book, and my life, have been enriched by my experiences with them.

My gratitude to Joel Kramer is profound. From my studies with him of hatha and jnana yoga I have learned an enormous amount about human awareness, the mind–body, and the deep structures of Eastern and Western psychological thought. His feedback on an earlier draft of the book was enormously helpful. Joel's life and teachings are a living embodiment of the integration of Eastern and Western ways of knowing and being. Our cross-country friendship has been of inestimable inspiration and importance to me.

I wish to thank my parents, Joyce and Harris Rubin, for creating a home congenial to intellectual and contemplative pursuits and for their ongoing love and support over the years. The origins of my contemplative and psychological interests were first nourished in this milieu.

Jerry Gold's ongoing interest and support have been deeply appreciated. I am grateful to Jerry for reading and commenting on several chapters of a penultimate draft and for the crucial role he played in this book being published.

In addition to reading an earlier version of this manuscript and sections from the penultimate draft, Alan Roland has provided intellectual and practical advice and guidance. From him I have become more attuned to the importance of sociocultural factors in thinking about psychoanalysis and Buddhism.

Dan Goleman, Mary Sheerin, Kendra Smith, and Denis Walsh read and offered enlightening feedback on an earlier version of this manuscript. I am grateful to them for their intellectual support, generosity, and guidance.

Jim Jones read the whole manuscript and provided invaluable intellectual dialogue, emotional support, and practical guidance. This work has been enriched in important ways because of his feedback. I am deeply grateful for his generosity.

A book can get sidetracked or stalled. A fortuitous encounter in the recent past with Victor Preller, with whom I had studied philosophy and religion at Princeton University over 20 years ago, played an important role in mobilizing me to finish this book at a crucial juncture.

Conversations over the years with friends and colleagues have also been enormously invaluable. Of those I remember, I would like to single out Diana Alstad, Ken Barnes, Gary Brill, Peter Cohen, Louise DeCosta, Doris Dlugacz, Bettina Edelstein, Mark Finn, Jerry Gargiulo, Vernon Gregson, David Kastan, Lenny Kriegel, Dorthy Levinson, JoAnn Magdoff, John McDargh, Stephen Mitchell, John Moody, Eugene Murphy, Karen Peoples, Michael Post, Victor Preller, Lobsang Rapgay, Chuck Simpkinson, Nancy Smart, Karen Smyers, Charles Spezzano, Tony Stern, John Suler, and Marc Wayne.

For invaluable, and sometimes emergency, computer assistance, I am grateful to Michael Powell, Bruce Rodin, and Carol Williams. Their generosity and support have been much appreciated.

Institutions, as well as individuals, have been essential to the genesis and completion of this book, providing an opportunity to share and refine various chapters. I am deeply grateful to the C.G. Jung Foundation for providing a forum over the years for the presentation of my ideas. My heartfelt thanks to Aryeh Maidenbaum and Janet Careswell who made it possible for me to teach Psychotherapy East and West at their institute. Their intellectual openness is inspiring. Several of the chapters in this book were presented in earlier incarnations at the Psychotherapy and the Spirit series of The Cafh Foundation, a nondenominational spiritual organization in New York City. I am grateful to Peter Cohen and his staff, who invited me to speak. I would also like to thank the organizers of the semiannual Buddhism and Psychotherapy conference for providing an opportunity to share my ideas. I am grateful to Mark Finn who made this possible. I am also very grateful to Marlin Brenner and his staff at the Institute for Training in Psychoanalytic Psychotherapy, who invited me to participate in the Healing the Suffering Self conference in 1994. I first presented an earlier draft of Chapter 4 at that time. I am also grateful to John Suler of Rider College, who invited me to speak at a conference he organized in Philadelphia in 1989 on Psychotherapy and Eastern Thought.

I am very grateful to Mariclaire Cloutier and Michele Fetterolf at Plenum Press for their patience, professionalism, and support in transforming my manuscript into a book.

The friendship, love, and support of Mary have been enormously sustaining. She has read this manuscript with an eye for style and has fostered and protected the tranquility and peace of mind that was necessary to complete this project. My emotional gratitude to her is profound.

Contents

Psychotherapy and Buddhism

Toward an Integration

Introduction

The contemporary world is in psychological crisis. There is a pervasive sense of powerlessness and purposelessness, alienation from self and others, enormous fear and suffering, and a dangerous rise in sadistic and self-destructive behavior. Nihilism, depression, addiction, and oppression have reached epidemic proportions. In such a psychological climate it is not an exaggeration to speak of selves-at-risk.

This occurs in a context of social disintegration. Traditional sources of social cohesion, meaning, and solace—family, religion, and politics—are eroding. Divorce is rampant, geographic mobility divides extended families, and corruption and rampant self-interest among political and religious leaders undermine faith in external solutions to our problems.

As the comforting, meaning-giving moral and cosmic horizons that had guided human choices and actions disappear, late-twentieth-century citizens experience ontological homelessness and psychological dislocation and chaos. Bereft of the orienting social and cosmic framework of action that helped people in the premodern world navigate their existence, the conduct of modern citizens becomes guided by individualistic and essentially self-centered preferences and wishes. This produces what Max Weber termed a "disenchantment" of the world, "flattened" and "narrowed" lives of "self-absorption" (Taylor, 1991, p. 4) and a sense of personal alienation.

In the face of the assault on the self that contemporary people confront, two opposing responses are discernible. On the one hand, there is a dramatic—and ultimately self-destructive—commitment to escape the burdens and responsibilities of being a human being through such means as immersion in dogmatic and divisive fundamentalist ideologies and cults and self-anesthetizing behavior (including compulsive work and consuming, drugs, and television) which increase self-blindness and self-mistrust and foster human alienation and suffering. On the other hand, there is an increased concern with expanding self-

1

awareness, defining and refining one's personal identity, and searching for self-fulfillment. There is a hunger for organizations, experiences, and practices that might provide meaning or lessen suffering.

Self-knowledge has traditionally been seen as crucial in increasing self-awareness, mitigating the power of self-unconsciousness and the irrational and improving the quality of human life. There have traditionally been three paths to self-knowledge. Rational introspection was utilized by philosophers from Socrates to Nietzsche. The second path, older than systematic introspection, was aimed at gaining such knowledge by altering states of consciousness, either with the aid of hallucinogenic drugs or by means of meditation. The third, and most recent strategy, is psychoanalysis (Bergmann, 1976, p. 3).

Attempts to understand human nature and conduct did not originate with nineteenth-century Western psychology. Western psychology, now about 100 years old, the product of European and American civilization, is only one of the latest efforts by humankind to understand itself.

While some of these prepsychoanalytic endeavors have been loose collections of folklore and myth, other societies have developed sophisticated theories of personality and methods of transformation and cure. Eastern religions, particularly Buddhism, provide one of the most fertile and well-formulated of these psychologies. Like their Western counterparts, these psychologies contain numerous theories and schools of practice. But while there are crucial conceptual and technical differences among these religions, there is far less divergence among the psychologies underlying them. Some of the commonalities are a concern for the immediate reality of one's experience, a critique of humans as they typically are, the positing of an ideal mode of being, a delineation of the forces that interfere with achieving optimal health, and penetrating techniques for self-transformation.

Classical Buddhism, which was the largest religion in the world until this century, offers arguably the most highly elaborated and comprehensive of these psychologies. Buddhism is the codification of Gautama Buddha's (536–438 BC) insights about human psychology developed in the course of his self-investigations utilizing meditative practices. Virtually every meditative system in the West (Zen, Transcendental meditation, and so forth) bears its stamp. It has been preserved, relatively unaltered, in the Theravadin Buddhist tradition.

For many centuries, Buddhism has been the dominant spiritual tradition in most parts of Southeast Asia, China, Korea, and Japan. It has had an enormous influence on the intellectual, cultural, and artistic life of these countries.

Buddhism, like psychoanalysis, is not a single doctrine.[1] In the 2500 years since its inception it has developed into different systems of theory and practice. The evolution of Buddhism resembles, in a certain sense, the banyon tree, with offshoots of the taproot (reinterpretations of Buddha's teachings) generating new branches that extend in various directions. At this point there is no monolithic theory of Asian psychology. Among the dozen or so schools that exist today, the most influential are the Theravadin Buddhists in Southeast Asia; the Ch'an school of China; Zen in Korea, Japan, and the United States; and the various sects of Tibetan Buddhism scattered throughout the world.

Theravadin Buddhism is the earliest and arguably the most psychologically oriented Buddhist school. Theravadin Buddhism is a nontheistic ethical and psychological training system that served as a prototype for subsequent Buddhist schools such as Zen and Tibetan Buddhism (Goleman, 1977). It has been well described in the classical (Buddhaghosa, 1976; Nyanaponika, 1972; Thera, 1941) and contemporary (Goldstein, 1976; Kornfield, 1977) meditative literature and has been largely preserved in its original form in contemporary Southeast Asia and the United States. The core practice of Theravadin Buddhism is Vipassana meditation, which cultivates nonjudgmental attentiveness to whatever is occurring in the present moment. It is claimed that this leads to heightened perceptual acuity and attentiveness, facilitates insight into the nature of psychic functioning, eradicates suffering, and promotes compassion, wisdom, and moral action.

Arnold Toynbee once suggested that future historians would conclude that the most significant event of our age was the introduction of Buddhism to the West. In recent years there has been a burgeoning interest in the relationship between the Buddhist and psychotherapeutic traditions among psychotherapists, researchers, and spiritual practitioners. The nascent dialogue between the Buddhist meditative and Western psychoanalytic traditions has been perhaps the most illuminating aspect of this historic encounter, yielding crucial insights about human development, conceptions of selfhood, psychopathology, and cure (Engler, 1983; Rubin, 1991, 1992).

---

[1]Psychoanalysis and Buddhism are not one thing. Since meaning, as Bakhtin (1986) knew, is the product of an interaction or dialogue between reader and text, rather than a monological essence waiting to be found in a neutral, fixed manuscript, there is no singular, settled or definitive Buddhism or psychoanalysis. "Buddhism" and "psychoanalysis" are heterogeneous and evolving: a multitude of beliefs, perspectives, and theories, cocreated and transformed by readers from different psychological, historical, sociocultural, and gendered perspectives.

The relationship between psychoanalysis and Buddhism has intrigued and perplexed me for the past 18 years. I have been deeply impressed by both traditions; significantly enough to spend years examining what they each might illuminate about the art of living and how they might facilitate personal transformation and social amelioration. In the years that I have been practicing psychoanalysis and meditation concurrently, my professional and personal lives have been immeasurably enriched. Participating in each discipline has continually enriched the practice of the other. Meditation practice deepened and refined my ability to listen with clarity and empathy to myself and my patients and lessened restrictive unconscious attachments. Psychoanalysis enhanced my capacity to understand the influence of the past on the present, especially unconscious allegiances to confining modes of being, that compromised the possibility of choice and freedom and contributed to unconscious psychological obstacles in spiritual practice. Practicing both disciplines has cultivated qualities and traits such as heightened awareness and insight, expanded empathy and deepened compassion, increased joy and reverence for life, and enhanced moral integrity, which could contribute to the creation of a more humane world.

Practicing psychoanalysis and meditation has led to turmoil as well as transformation. Each discipline has seemingly antithetical viewpoints on topics ranging from the cause of psychological suffering to the method of treatment. How could both systems be coherent, valid, and effective if they gave opposing answers to questions about such topics as the nature of the self, what is optimum psychological functioning, and what facilitates and what interferes with change?

Various questions arose as I grappled with these issues: Are psychoanalysis and meditation different approaches to the same goal? Or are they different approaches to different goals? Could meditation heal emotional wounds and facilitate psychological well-being? Could psychoanalysis foster an increased sense of compassion and morality and assist one's meditation practice? What are the limitations and contributions of both systems as paths to liberation?

Dialogues with teachers and practitioners of both disciplines and perusal of the scant literature on the relationship between psychoanalysis and meditation did not help me reconcile how these two transformative disciplines could be integrated. Despite the historical and linguistic connection between psychology and spirituality (the word *therapeutae* from which therapy derives refers to a healing sect of mystics residing in Egypt in the first century AD), mental health professionals and spiritual practitioners seemed, for the most part, to either

(1) keep psychoanalysis and Buddhism segregated; (2) romanticize one and devalue the other; or (3) recommend an "integration" that in actual practice subordinated one to the other. When Eurocentrism flourishes, by which I mean the view that European standards and ideals are the center of the universe, then psychoanalysis and Western psychotherapies tend to be deified and Buddhism is denigrated. When what I have recently termed "Orientocentrism" reigns, which refers to the tendency to treat Oriental culture as sacred, then Buddhism is uncritically over-valued and psychoanalysis is disparaged or neglected (cf. Rubin, 1993).

Psychoanalysts generally ignored or devalued Buddhism and medi-tation, while Buddhists usually neglected psychoanalysis. Western mental health professionals who valued meditation tended to reject a psychoanalysis that they had not accurately assimilated, presenting distorted and reductionistic accounts of it based on antiquated and superseded psychoanalytic theories, e.g., Freudian drive theory. The enormous potential of psychoanalysis to facilitate growth and transfor-mation was obscured by accounts of it by Buddhist-oriented authors, which failed to recognize recent theoretical and clinical innovations in understanding of psychic structure, human motivation, and the curative aspects of the therapeutic process offered in psychoanalytic object rela-tions theory, interpersonal psychoanalysis, and self psychology.

My conflict involved the clash between my ongoing *experience* that both psychoanalysis and Buddhism were enormously valuable and the pervasive tendency in the literature to preclude their actual encounter. Whereas meditators and analysts maintained that there was a clear demarcation between the Buddhist path of self-transformation and the psychoanalytic path of self-integration, my experience suggested that they overlapped and were interwoven. Psychoanalysis, for example, often facilitated self-transformation, while meditation frequently en-hanced self-integration.

At first glance it might seem that speaking of psychoanalysis and Buddhism in tandem is advocating a forced and unproductive associa-tion. After all, several fundamental disparities between them exist. Bud-dhism is a spiritual system developed 2500 years ago in India for attain-ing enlightenment; psychoanalysis is a psychotherapeutic system that arose in Europe in the late nineteenth century to address psychopathol-ogy and mental illness. To attain enlightenment, Buddhism recom-mends recognizing the illusoriness of our taken-for-granted quotidian sense of self as a unified, static, unchanging, autonomous entity; most analysts, with the exception of Lacanians, claim that strengthening the self is crucial to the psychoanalytic process. Buddhism emphasizes the necessity of letting go of all desires and self-centeredness, while psychoanalytic self psychology maintains that ideals and goals play a crucial role in psychological well-being.

Skepticism about the possibility of integration is complicated by the fact that there are also similarities between both traditions that make comparing them intriguing. Each is concerned with the nature and alleviation of human suffering and each has both a diagnosis and "treatment plan" for alleviating human misery. The three other important things they share make a comparison between them possible and potentially productive. First, they are pursued within the crucible of an emotionally intimate relationship between either an analyst and analysand or a teacher and student. Second, they emphasize some similar experiential processes: evenly hovering attention—free association and meditation. Third, they recognize that obstacles impede the attempt to facilitate change, e.g., resistance—defensive processes in psychoanalysis and "hindrances"–"fetters"–"impediments" in Buddhism.

It is now an auspicious time for cross-cultural communication and dialogue. The physical and conceptual walls dividing "East" and "West" are crumbling. Opportunities for intercultural dialogue are enormous.

With their exemplary tools for self-investigation and their enormous resources for exploring human self-blindness and facilitating self-awareness, psychoanalysis and Buddhist meditation have something crucial to contribute to the difficulties and demands of modern living.

Psychoanalysis and Buddhism also offer fertile possibilities for cross-pollination. Mutual enrichment, however, has been impeded by the restrictive perspective of previous studies that have tended to adopt either a Eurocentric or an Orientocentric perspective. In the next chapter I will attempt to elucidate how this is true of even those viewpoints that attempt to create a complementary synthesis of "Eastern" and "Western" thought, with the exception of the work of Kramer and Alstad (1993).

This work attempts to sketch an alternative to the Eurocentrism and Orientocentrism that has plagued the field; a close encounter of a new kind. Neither psychoanalysis nor Buddhism are absolute truths, a set of theories or procedures to be followed in every situation. Both traditions are not God-like, timeless truths, but human creations developed by particular people at specific historical moments in order to solve certain historically distinctive problems and serve particular human interests.

It will be the argument of this book, to draw from *Hamlet*, that "there are more things in heaven and earth ... than are dreamt of" in either psychoanalytic or Buddhist psychologies alone, and that both psychoanalysis and Buddhism can both enormously benefit from an egalitarian dialogue characterized by mutual respect, the recognition of differences, and a genuine interest in what they could offer each other. This relationship is not without disagreements, points of contention, and conflict. Such turmoil can be healthy insofar as it impedes orthodoxy, dogmatism, and premature closure and can promote cross-pollination and growth.

I shall be studying Buddhism and psychoanalysis in tandem, not in order to discover which is Really True or superior, but to encourage a dialogue in which some of the blindspots and hidden strengths of each tradition might emerge. Possibilities that are not available when each tradition is explored in isolation would then become more readily apparent and each system would be more efficacious than if pursued alone.

My perspective has been shaped—sometimes explicitly and sometimes implicitly—by a Taoist and Jungian respect for the unity and value of apparent opposites, such as self and other, connectedness and "nonattachment." Valuing both sides of what is normally polarized fosters an inclusive rather than a dichotomous viewpoint on psychoanalysis and Buddhism and on living. I see each pair as different and vital, necessary, and by itself an incomplete perspective. Each facet of the polarity thus needs to be integrated with its apparent opposite. We need to acknowledge the other as well as the self, immanence as well as transcendence, psychology no less than spirituality.

My perspective is dialogic, not dialectic. The contact of opposites leads not to a synthesis in which either is superseded, but to a *conversation* in which both have something novel to communicate and in which something new is created.

My stance of "bifocality" or "reciprocity of perspectives" (Fischer, 1986, p. 199), in which psychoanalysis and Buddhism can mutually question and challenge each other, encourages a placing of taken-for-granted assumptions in the foreground, a relativizing of apparent eternal truths, a highlighting of theoretical blindspots, and an illuminating of unsuspected possibilities.

The perspective of cultural hybridization that underlies my work, of valuing and being inside and outside two traditions, is more difficult than simply theorizing from within one tradition. I am not unaware that there may be both psychoanalysts and Buddhists who may feel that psychoanalysis or Buddhism alone is a sufficient antidote to the collective ills that face us. The historical record leaves me deeply unconvinced. From my perspective the importance of challenging the hegemony of the Eurocentrism and the Orientocentrism that has permeated the field of East—West studies *and* the possibilities that an egalitarian perspective might generate warrant the risk of such an interdisciplinary investigation.

What I have discovered since approaching psychoanalysis and Buddhism in this way is that both traditions have a great deal of merit but neither provides a complete picture of human nature, transformation, and liberation. Each offers a valuable and incomplete perspective, neglecting indispensable elements included in the other. For example, Buddhist models of health could teach psychoanalysis that there are possibilities for emotional well-being that far exceed the limits described by psychoanalytic models, while psychoanalysis could help Buddhists understand some of the unconscious interferences to meditation practice and the growth process. Buddhists could teach psychoanalysts about states of dereified, decommodified, and non-self-centric subjectivity. Psychoanalysis could teach Buddhists about the recurrent, unconscious psychological conditioning and restrictive patterns of relating to others arising from one's past that shape and delimit human life. It could also elucidate the psychological dangers of neglecting human agency.

Since neither tradition has the last word on these issues, both traditions could be enriched if their respective insights were integrated into a more inclusive and encompassing perspective, which currently does not exist, that takes into account their respective contributions and elucidates their blindspots, while attempting to bolster their limitations.[2]

Once it is recognized that both traditions are valuable and incomplete, two questions emerge: What does each tradition illuminate? What does each tradition omit? With these two questions in mind, I shall examine psychoanalysis and Buddhism along three dimensions common to any psychological, religious, or philosophical system: their (1) view of human nature; (2) model of ideal health; and (3) conception of

---

[2]Cross-fertilization between psychoanalysis and Buddhism is not a panacea. Problems remain for each discipline even after contact with the other. In stressing the value of a complementary perspective, I do not mean to suggest that either psychoanalysis or Buddhism will be without flaws after they have filled in some of each other's gaps and omissions. There are shortcomings in each that the other does not remedy. Psychoanalysis, for example, may not develop greater sensitivity to the roles cultural, socioeconomic, and historic forces play in human development after being exposed to Buddhism. Buddhism may not become more attuned to issues of gender and the oppression of women through its encounter with analysis. My exploration of certain ways psychoanalytic and Buddhist theories and practices complement and enrich each other does not eclipse the necessity for further transformation of each discipline. This topic is beyond the scope of this book.

the process designed to reach its stated goals, which includes a theory of the obstacles to the process (cf. Shapiro, 1989). This volume is divided into three sections based on a tripartite demarcation: psychoanalytic and Buddhist views of self (Part I), health (Part II), and the process or the path designed to promote the health of the self (Part III). The two introductory chapters will provide a context for the study as a whole by presenting an overview of Buddhist and psychoanalytic history and theory and a review both of how others have conceived of their possible relationship and of how I think about them.

I believe that approaching psychoanalysis and Buddhism from a multidimensional perspective will encourage the emergence of the emancipatory dimensions of both traditions, helping them become significantly richer than if studied alone. The union of the work of the Indian sage meditating in the middle of the first millennium before Christ and the Viennese doctor who explored the traumas and dreams of his patients in Europe in the first three decades of the twentieth century could reduce self-blindness, encourage an expanded view of the possibilities of human functioning, facilitate a deepened understanding of how to investigate the self and work with unconscious barriers to change, and offer a method of treatment that can facilitate profound healing and transformation. The contemplative psychoanalysis or analytic meditation that such an encounter might foster could offer "resources of hope" (Williams, 1989) for postmodern selves-at-risk.

In *Analysis Terminable and Interminable*, Freud (1937) recommended that analysts return to analysis every 5 years. He was acknowledging the unmasterability of mind, the way unconsciousness never ends. Since life, unlike writing, as Robert Lowell (1977) instructs us, never finishes, the perspectives in this volume are vulnerable to challenge and revision in the face of different questions and new knowledge and insights. It could not be otherwise. I thus hope that what I have asserted in this work is treated as footprints of my journey rather than a blueprint for yours.

REFERENCES

Bakhtin, M. (1986). *Speech genres and other late essays.* Austin: University of Texas Press.

Bergmann, M. (1976). What is psychoanalysis. In M. Bergmann & F. Hartmann (Eds.), *The evolution of psychoanalytic technique* (pp. 2–16). New York: Basic Books.

Buddhaghosa. (1976). *The path of purification* (B. Nyanamoli, Trans.). Berkeley, CA: Shambhala.

10 *Introduction*

Engler, J. (1983). Vicissitudes of the self according to psychoanalysis and Buddhism: A model of object relations development. *Psychoanalysis and Contemporary Thought, 6*, 29–72.
Fischer, M. (1986). Ethnicity and the post-modern arts of memory. In J. Clifford & G.E. Marcus (Eds.), *Writing culture, the poetics and politics of ethnography* (pp. 194–233). Berkeley: University of California Press.
Freud, S. (1937). Analysis terminable and interminable. In J. Strachey (Ed. & Trans.), *The standard edition of the complete psychological works of Sigmund Freud*, Vol. 23 (pp. 216–253). London: Hogarth Press.
Goldstein, J. (1976). *The experience of insight: A natural unfolding.* Santa Cruz, CA: Unity Press.
Goleman, D. (1977). *The varieties of the meditative experience.* New York: Dutton.
Kohut, H. (1977). *The restoration of the self.* New York: International Universities Press.
Kornfield, J. (1977). *Living Buddhist masters.* Santa Cruz, CA: Unity Press.
Kramer, J., & Alstad, D. (1993). *The guru papers: Masks of authoritarian power.* Berkeley, CA: North Atlantic Press.
Lowell, R. (1977). *Selected poems.* New York: Farrar, Straus & Giroux.
Nyanaponika, T. (1962). *The heart of Buddhist meditation.* New York: Samuel Weiser.
Nyanaponika, T. (1972). *The power of mindfulness.* San Francisco: Unity Press.
Rubin, J.B. (1991). The clinical integration of Buddhist meditation and psychoanalysis. *Journal of Integrative and Eclectic Psychotherapy, 10*(2), 173–181.
Rubin, J.B. (1992). Psychoanalytic treatment with a Buddhist meditator. In M. Finn & J. Gartner (Eds.), *Object relations theory and religious experience* (pp. 87–107). Westport, CT: Praeger.
Rubin, J.B. (1993). Psychoanalysis and Buddhism: Toward an integration. In G. Stricker & J. Gold (Eds.), *Comprehensive handbook of psychotherapy integration* (pp. 249–266). New York: Plenum Press.
Shapiro, D. (1989). Judaism as a journey of transformation: Consciousness, behavior, and society. *Journal of Transpersonal Psychology, 21*(1), 13–59.
Taylor, C. (1991). *The ethics of authenticity.* Cambridge, MA: Harvard University Press.
Thera, S. (1941). *The way of mindfulness: The Satipatthana Sutra and commentary.* Kandy, Sri Lanka: Buddhist Publication Society.
Williams, R. (1989). *Resources of hope.* London: Verso.

1

Psychoanalytic and Buddhist History and Theory

HISTORICAL SETTING

Buddhism arose in the sixth century BC in India. This epoch was a watershed in the intellectual and spiritual development of the world. The Greek rationalist philosophers, the Jewish prophets, Confucius in China, and Buddha in India all illuminated human history during this time.

The history of India is shrouded in obscurity. The archeological record is incomplete. Buddhism's fundamental "ahistoricity," in which time and place become irrelevant to those who seek to dwell in the presence of the infinite, make historical reconstruction difficult (McNeill, 1963).

Early Buddhist works suggest traces of aristocratic and republican polities comparable to the Greek city–states. But in the Ganges valley, centralized monarchical states began to dominate the scene from about the eighth century BC. By about the sixth century BC, these monarchies were either absorbed or had established their control over most of the small republican states of northern India.

From all accounts, the India of Buddha's time was in transition and ferment. There was "rapid social change.... So hitherto significant ordered collectivities, and the individuals within them, were no longer able to construe or guide their own fate" (Carrithers, 1985, p. 254). The local tribal republican government and aristocratic society were disrupted by new social and political forces. As territorial subordination replaced tribal bonds and agriculture came to replace cattle tending, the old tribal solidarity was compromised. The aristocratic style of life, expressed politically in the tribal republics, crumbled under the pressure of urbanism and centralized monarchy. In a world in which traditional social structures had vanished, there were those who probably joined the victors and became servants of the ascending monarchies, while others may have sought a spiritual resolution to their worldly frustration, joining religious orders. It has been suggested that the breakup of the local tribal units and their replacement by kingdoms eroded ethnic ties and the sense of security they fostered, which led to fundamental psychological turmoil.

It was a time of enormous suffering and narcissism (cf. Roccasalvo, 1982). Dissatisfaction was prevalent. There was a "great wave of pessimism" (DeBary, 1972, p. 5). The sources were varied. Daily life was enormously difficult and unappealing. "Life in the home," noted Buddha, "is cramped and dirty" (quoted in Carrithers, 1983, p. 21). Drought, disease, and hunger were ubiquitous menaces. Every year, floods destroyed hard-earned crops wrung from the earth, monsoons spawned droughts, famine, dysentery, cholera, and countless other ills, weakening people for the predatory beasts (cf. Stryk, 1968, p. xxxiii).

That narcissism may have been prevalent in Buddha's India and was a concern of Buddhism is suggested by the frequent references in the Pali Buddhist texts to two narcissistic conceptions of human beings held by the "uneducated manyfolk" and Buddha's contemporaries. A central facet of Hinduism, which was the predominant philosophical and religious perspective in India prior to Buddhism, was the belief in *Atman*, or an unchanging self. Buddhism was, in certain fundamental ways, a protestation against certain aspects of Hinduism, including its emphasis on the primacy of the self. Buddha unequivocally opposed Eternalism or self-immortality, the view that the soul is "permanent, steadfast, eternal, and not subject to change" (Roccasalvo, 1982, p. 209), and the equally self-preoccupied Annihilationism, the view that the body is the "final locus for utter and complete dissolution" (Roccasalvo, 1982, p. 208). These positions are illustrated by Buddha's descriptions of two types of teachers:

> Here a certain teacher sets out soul as something real and permanent in the present life as well as in the future life. Again, another teacher sets out soul as something real and permanent as far as this world is concerned but does not say so with regard to any future existence.... The teacher of the first order is to be understood as a teacher who upholds the doctrine of Eternalism ... the teacher of the second order ... the doctrine of Annihilationism ... (Roccasalvo, 1982, p. 210)

The Buddhist doctrine of *anatta*, the essential selflessness of humans, which served as the cornerstone of Buddhist teachings, opposed these two narcissistic theories.

Because of the pervasiveness of suffering, Indians had a "passionate desire for escape, for unison with something that lay beyond the dreary cycle of birth and death and rebirth, for timeless being, in place of transitory and therefore unsatisfactory existence" (Basham, 1988, pp. 44–45). Escape from one's condition in life was virtually nonexistent because India was ruled by the estates system. The estates prescribed an orderly, preordained hierarchical relationship between people, with each person having certain responsibilities toward others. The estates defined a person's position in life. The world of India was hierarchically ordered in a prescribed manner: those who labor, those who fight, and those who pray. The servants, who occupied the lowest position, were ineligible for the benefits of the sacrificial religion and were compelled to a life of servitude by the three other estates. The commoners were the producers. The Warriors, whose duty it was to rule, fight, and pray for sacrifice, wielded power. The Brahmans were the priests of the sacrificial religion and the intellectuals.

Buddhism presented a fundamental and plausible refutation of the estates theory. Success and fate are the result of good deeds. The good deeds of the poor will garner their just desserts in the next life, while those who abuse power will be punished. Whether one was destitute or favorably endowed by birth, successful in business or defeated in battle, this theory could explain it. This offered Indians a release from the squalid exigencies of daily existence.

BUDDHISM

Siddhartha Gautama, the Buddha, was born around 536 BC in a small province in northern India on the border of present-day Nepal. His title—the Buddha—became his message. The Sanskrit root *Budh* connotes both to wake up and to know. Buddha means the "Awakened One" or the "Enlightened One." While the vast majority of humans lived as if they were emotionally "asleep," unable to escape from stifling patterns and unending suffering, Buddha roused himself from psychological slumber. Buddhism is the ethical psychology based on his discoveries.

Legend has it that before he was born his father, Suddhodana, the head of the Shakya clan and ruler of the principality of Kapilavastu, received a disturbing prophecy about his future son. A choice between two diametrically opposed destinies lay before the rajah's heir: He might become a great sovereign or a famous ascetic. Fearing the latter, Gautama's father attempted to prevent him, at all costs, from witnessing misery or unhappiness. Gautama lived amid great luxury and was shielded from exposure to sickness, old age, and death.

In his 20s he disobeyed his father's injunctions and left the sheltered palace compound. In the streets of Kapilavastu he encountered four sights: an old man, a corpse, a sick man, and a holy mendicant—known in Buddhism as the "four signs"—which fundamentally transformed the direction of his life. He was bewildered and horrified by the realization that old age, sickness, and death were the common fate of humankind. Before returning to the palace, he saw a peaceful wandering ascetic. In the serenity of this recluse, Gautama sensed the only response to his growing disillusionment.

After witnessing these three disturbing signs of human suffering and mortality, he returned to the splendor of his home. He fell asleep. When he awoke, his attendants, who usually entertained him with dance and song, were asleep. As he observed their "bodies wet with trickling phlegm and spittle; some grinding their teeth, and muttering and talking in their sleep" (Warren, 1977, p. 61), he was filled with aversion. His home "began to seem like a cemetery filled with dead bodies impaled and left to rot" (Warren, 1977, p. 61). He felt that "life in the home is cramped and dirty, while the life gone forth into homelessness is wide open" (Carrithers, 1983, p. 21). He said out loud to himself: "how oppressive and stifling is it all.... It behooves me to go forth on the Great Retirement this very day" (Warren, 1977, p. 61).

Once he had perceived the inevitability of illness, bodily pain, and mortality, he could not return to the normal pleasures of worldly pursuits. He said,

> Why, since I am myself subject to birth, ageing, disease, death, sorrow and defilement, do I seek after what is also subject to these things. Suppose being myself subject to these things, seeing danger in them, I were to seek the unborn, undiseased, deathless, sorrowless, undefiled, supreme surcease of bondage, the extinction of all these troubles. (Carrithers, 1983, p. 21)

Since he felt life was inevitably painful and enslaving, renouncing worldly life seemed, in his view, the only way to escape human bondage. His encounter with old age, sickness, and death was so troubling that he decided to forsake his life of ease, leave his home, his wife, and his children that very evening and become a spiritual seeker.

His quest lasted 6 years and exposed him to a variety of spiritual teachings and practices. It is said that his journey occurred in three phases. First, he studied yoga and philosophy with two Hindu masters who propounded ascetic and sensualistic practices. But he did not find what he sought; neither pursuing self-mortification nor self-indulgence released him from suffering. Recognizing that neither method extinguished the flame of desire or led to the liberation he sought, he devoted the final phase of his search to religious contemplation and meditation.

He decided to explore a more moderate path involving intensive self-scrutiny in the hopes of "destroying passions net" (Warren, 1977, p. 76). One evening he sat under a tree in a lotus position in northern India, south of the town now known as Patna, vowing that he would not move until he solved the vexing problems that besieged him.

After 6 days, it is said that his eyes opened on the rising morning star and he experienced a profound clarification of his searching; an understanding of the riddle of human existence; a freedom from crippling psychological illusions and a vision of the path to eradicating human suffering and attaining freedom.

His self-investigations led to what he termed an "exalted calm" and a blissful self-emancipation. He said:

> Through birth and rebirth's endless round,
> Seeking in vain, I hastened on,
> To find who framed this edifice
> What misery!—birth incessantly!
> O builder! I've discovered thee!
> This fabric thou shalt ne'er rebuild!
> The rafters all are broken now,
> And pointed roof demolished lies!
> This mind has demolition reached,
> And seen the last of all desire. (Warren, 1977, p. 82)

After this experience he eventually returned to the quotidian world and became a religious teacher. He taught that suffering pervades human existence and is caused by one's attachment and clinging to an illusory belief in the notion that there is an independent, abiding self. He claimed that there is no self and there is no inner director in control. Psychic reality, in his view, is created by what we think, not by a self or an external world: "the mental natures are the result of what we have thought, are chieftained by our thoughts, are made up of our thoughts" (Buddha, 1950, p. 58).

Through purity of thought and deed he maintained that it is possible to escape the tormenting cycle of rebirth with its unending suffering and inevitable death. Unwholesome thinking usually leads to unwholesome actions that create negative consequences such as creating another life and body and causing one to be reborn: " 'to be born here and die here, and die here and be born elsewhere, to be born there and die there, to die there and be born elsewhere ... this is the round of existence.... He that still has the corruptions is born into another existence; he that no

longer has the corruptions is not born into another existence.' " Awareness, Buddha, maintained, is the path to the *deathless*: "Vigilance is the
abode of eternal life, thoughtlessness is the abode of death. Those who
are vigilant (who are given to reflection) do not die" (Buddha, 1950, p. 66).

Buddha's teachings can best be understood against the background
of Hinduism of which they arose. Buddhism was, as I briefly alluded to
earlier, an Indian Protestantism in the original sense of witnessing (*tes-
tis*) for (*pro*), and in the more recent connotation of protesting against
something else. Buddhism began as a revolt against six aspects of Hinduism: authority and tradition, ritual, speculation, grace, mystery, and a
personal God. Buddha felt each had gotten out of hand (cf. Smith, 1986).

Slavish adherence to authority and tradition had justified and perpetuated the privilege and dominance of the ruling Brahman class.
Spiritless performance of rituals and preoccupation with metaphysical
questions had become a sterile substitute for authentic religious experience. Concepts of divine sovereignty and grace had promoted passivity
rather than spirituality. Religious mystery had degenerated into religious mystification.

Onto this sterile religious stage Buddha emerged and forged a religion devoid of each of these six elements. His attack on authority and
tradition was unequivocal: "do not accept what you hear by report, do
not accept tradition, do not accept a statement because it is found in our
books, nor because it is in accord with your beliefs, nor because it is the
saying of your teacher.... Be a lamp unto yourselves" (Burtt, 1955, pp.
49–50).

Ritual and speculation fared no better. Buddha advocated a religion
without ritual. In fact, "belief in the efficacy of rites and ceremonies" is
one of the ten fetters, or obstacles to spiritual practice in classical
Buddhism. Buddha was not uninterested in metaphysical questions—
he had given them close attention—but he felt that "greed for views" on
such questions tended "not toward edification" (Burtt, 1952, p. 15) and
detracted from the crucial concerns of reducing human misery. There is
some evidence that when abstract inquiries were put to him, he remained silent or directed the interrogator toward the more important
subject of how best to lead one's life.

Buddha's concerns were pragmatic and therapeutic. His primary
focus was psychological and ethical. He was more interested in alleviating human suffering than in satisfying human curiosity about the origin
of the universe or the nature of divinity.

For Buddha, neither God's grace nor divine intervention could aid
in this endeavor. He condemned all forms of supernatural divination

and soothsaying. The fatalism and passivity that God's sovereignty and grace often fostered was replaced by an encouragement to intense personal effort. He took pains to emphasize to his followers that none of them was to look upon him as a divine savior; that he only pointed out the path to freedom; that they had to "work out their salvation with diligence" (Burtt, 1952, p. 49).

His central teaching was the Four Noble Truths. This doctrine delineates the symptom, diagnosis, prognosis, and treatment plan for addressing human suffering.

The first Noble Truth presents the salient characteristic of human life, *Dukkha*, a Sanskrit word for awryness, unsatisfactoriness, and suffering. It refers to an "axle which is off-center with respect to its wheel" and to a "bone which has slipped out of its socket" (Smith, 1986, p. 150). Life, according to Buddha, is dislocated, out of joint, and full of suffering.

According to Buddha there are three types of suffering. There is the ordinary suffering of old age, sickness, and death. This is similar to what Freud (1895) referred to as the great inevitabilities of fate. Then there is the suffering caused by change. Change—personal, relational, environmental, occupational—can be unsettling. The third type of suffering parts company with Western psychological understandings. Buddha maintains that all conditioned states of mind inevitably lead to suffering. By becoming attached to what changes, humans, according to Buddhism, sow the seeds of their own suffering.

The second Noble Truth presents the cause of suffering: desire, attachment, and craving. There are three types of desire: desire for sense gratification, existence or nonexistence, and the clinging to self. The first type of desire corresponds to Freudian perspectives on pleasure principle functioning. The second relates to issues familiar to psychoanalysts treating patients suffering from issues related to the continuity of the self and selfhood in relation to others. The third type of desire in the Buddhist scheme is similar to narcissistic issues in psychoanalysis.

A brief synopsis of the Buddhist model of the mind helps place the second Noble Truth in perspective. Mind, according to Buddhism, is composed of three elements: (1) consciousness of (2) one or more of the five senses—seeing, hearing, tasting, touching, smelling—plus thinking (which in Buddhism is considered a sixth sense) and (3) a reaction of attachment, aversion, or impartiality to whichever of the six facets of experience one is aware of. Reading this paragraph, for example, there is either aversion, affection, or neutrality toward the thoughts that are arising. In this impersonal theory of mind, every instant of seeing, hearing, tasting, touching, smelling, or thinking is responded to with pleasure, unpleasure, or neutrality, but without anyone *having* those experiences.

Suffering, according to the Buddhist account, derives from our difficulty acknowledging a fundamental aspect of life: that everything is impermanent and transitory. Suffering arises when we resist the flow of life and cling to things, events, people, and ideas as permanent. The doctrine of impermanence also includes the notion that there is no single self that is the subject of our changing experience.

The third Noble Truth is that suffering can be eradicated. It is possible, according to Buddhism, to extricate oneself from psychological imprisonment and to reach a state of complete awakening or liberation called Nirvana, which means "to blow out" or "to extinguish." What is extinguished is personal desire. In this state, grasping and suffering have disappeared and the oneness of all life is evident. There is no equivalent in the history of Western psychology. "Health" in Western psychology, whether Maslow's self-actualization or the fully analyzed patient of psychoanalysis, is an arrested state of development according to Buddhism.

The fourth Noble Truth provides the map of how to experience enlightenment: the Noble Eightfold Path, which comprises right understanding or accurate awareness into the reality of life; right thought or aspiration; right speech, speaking truthfully and compassionately; right action, abstaining from killing, lying, stealing, adultery, and misuse of intoxicants; right livelihood, engaging in occupations that promote, rather than harm life; right effort, or the balanced effort to be aware; right mindfulness, seeing things as they are; and right concentration, or meditative attentiveness.

The Theravadin ideal of spiritual development was Nirvana or complete awakening and liberation. The *Arahant*, or one "worthy of praise" for conquering the enemies of awareness and wisdom, that is, greed, hatred, and delusion, attained this state and was believed to have escaped from reincarnation. The *Arahant* will be discussed in greater detail in Chapter 4.

None of Buddha's teachings were recorded during his lifetime. In the first few centuries after his death, several Great Councils were held by the leading members of the Buddhist order at which time the entire Buddhist teachings were recited aloud and interpretative disputes were addressed.

During the rainy season after Buddha's death it is said that 500 of his leading disciples convened the First Great Council. Ananda, Buddha's attendant, repeated all of the sutras or sermons and discourses; Upali, recited the *Vinaya*, the 250 rules of morality and discipline; and

Mahakashyapa presented the Abhidharma, the higher philosophical and psychological teachings.

Buddhism, like psychoanalysis, developed partisan schools and schisms. At the Second Great Council a schism developed regarding how strictly to follow the Vinaya rules. Ten thousand monks were expelled from the Council. They formed a school called the Mahasanghika, which flourished in Northern India. The remaining Buddhists, the Theravadins, or the school of the elders, Buddha's contemporaries (often erroneously known as the Hinayana, or the "small vehicle"), banded together in the south of India. The Theravadins continue to this day in Southeast Asia and parts of India and the United States. The Mahayana or "great vehicle" spread to the north and east and was eventually transplanted to Korea, Japan, Nepal, China, Tibet, and the United States.

Each school was termed a *yana*, or raft, that carried Buddhist practitioners across the sea of life to the shore of enlightenment. The words "small" and "great" vehicle refer to the respective restrictions and latitude of interpretation and practice of Buddhist doctrine. The Mahayanists, who do not object to being designated as the large vehicle, adhered less closely to early Buddhist teachings. The Hinayanists, who prefer to characterize their brand of Buddhism as Theravada, or the Way of the Elders, observed more of the letter of Buddha's teachings. In doing this, they claim to represent the original unadultered Buddhist teachings as taught by Buddha. They maintain that spiritual practitioners are "on their own" with progress being based on one's own efforts. Emancipation is not contingent on the salvation of others. No God or intercessory powers are available (Ross, 1966).

Mahayana Buddhism encompasses doctrines ranging from religious faith in the teachings of Buddha to elaborate philosophies and complex cosmologies. Unlike the Theravadins, Mahayanists believed in a personal God and a divine savior. For the Mahayanists, unlike the Theravadins, emancipation is contingent on the salvation of others. Grace and love are the sine qua non of the path.

For the Mahayanists, the Theravadin ideal was selfish. The proper focus of spiritual life should be on refraining from entering Nirvana in order to help others ascend the ladder of reincarnation and escape from the suffering of existence. The Theravadin ideal of spiritual development—the *Arahant*—was replaced with the Bodhisattva, "one whose essence (*sattva*) is perfected wisdom" (*bodhi*), a being who forsakes the quest for enlightenment for him or herself alone, but has vowed to help all other beings attain enlightenment. Bodhisattvas were believed to inhabit a heaven of their own making. Pointing to its doctrine of grace and its wider accessibility for laypeople, Mahayana Buddhism claimed to be the larger vehicle of the two. Theravadin Buddhism remains a unified tradition, while Mahayana has splintered into five schools that stress such elements as faith, intellectual study, reciting ritual formulas, and intuitive understanding.

In the past a great deal of attention in the West has been devoted to Mahayana Buddhism. Theravada Buddhism has usually been mentioned in terms of early Buddhist history and scriptures. Yet, Theravada monks and nuns and millions of lay disciples in Southeast Asia form the largest living Buddhist tradition in the world (Kornfield, 1977). The practices taught by the Theravadin Buddhist meditation masters in Thailand, Burma, and Sri Lanka are based on the original Pali scripture and subsequent transmission. It is this tradition that my study is primarily based on. References to Zen and Tibetan Buddhism, however, will appear throughout the text.

My choice of focusing on one tradition should not be construed as a judgment on the other schools of Buddhism. Different schools are different vehicles. Buddha's reflections bear repeating:

> Would he be a clever man if out of gratitude for the raft that has carried him across the stream to the other shore, he should cling to it, take it on his back, and walk about with the weight of it? Would not the clever man be the one who left the raft (of no use to him any longer) to the current stream, and walked ahead without turning back to look at it? Is it simply a tool to be cast away and forsaken once it has served the purpose for which it was made? In the same way the vehicle of the doctrine is to be cast away and forsaken once the shore of Enlightenment has been attained. (quoted in Smith, 1986, pp. 209–210)

As he was dying, Buddha is reported to have said to his attendant, Ananda: "Be a lamp unto thyself; pursue your deliverance with diligence" (cf. Burtt, 1955, p. 49). No one was selected to teach or govern the Buddhist community that outlived him. The Dharma, the teachings of the truth of how things are, not a person or institution, would be the teacher (Kornfield, 1977). In the next section I will present an overview of psychoanalysis.

PSYCHOANALYSIS

Four character ideals have vied for center stage in the history of Western life, according to sociologist Phillip Rieff (1963): the "political" subject of classical antiquity who participates in public life; the "economic" subject who retreats into a search for private fulfillment while enjoying the fruits of citizenship; the Hebraic and Christian "religious" person who substitutes faith for reason; and the late nineteenth- and twentieth-century "psychological" subject, who eschews any redemptive external doctrine or creed, whether political or religious, and attends to the workings of his or her own private universe of thoughts, feelings, dreams, and symptoms.

Psychoanalysis arose from the soil of the modern period in which there was a "despiritualization" of subjective reality (cf. Kovel, 1991), by which I mean a devaluation, marginalization, and pathologization of the spiritual. Spirituality does not flourish in a world in which science rather than religion is viewed as the ultimate arbiter of the nature of reality.

The "psychological" subject monopolized the stage of intellectual life in the West during the formation and development of psychoanalysis. "In the age of psychological man, the self," notes Rieff (1963), "is the only god-term" (p. 23). Selfhood, according to Baumeister (1987), became a problem in the modern period:

> During the Victorian era (roughly 1830–1900), there were crises with crises with regard to ... four problems of selfhood ... how identity is actively or creatively defined by the person, what is the nature of the relationship between the individual and society, how does the person understand his or her potential and then fulfill it, and how and how well do persons know themselves.... [E]arly in the 20th century, themes of alienation and devaluation of selfhood indicated concern over the individual's helpless dependency on society. (p. 163)

For psychological man, self-maximization, not participation in the polis, is the chief vocation. Interest in the workings of one's psyche replaces commitment to the life of the commons. Psychotherapeutic concern for the meaning of symptoms replaces questions about meaning or "ultimate concern." Better living, not the Good Life, becomes the main psychoanalytic preoccupation (Rieff, 1963).

Psychoanalytic assumptions about the self, whether classical or contemporary, are underwritten by Western values, particularly the "Northern European/North American cultural values and philosophical assumptions involving individualism" (Roland, 1995, p. 4). The individual in individualism is sacred: "the supreme value in and of itself, with each having his or her own rights and obligations.... Society is considered to be essentially subordinate to the needs of individuals, who are all governed by their own self-interest ... (Roland, 1995, p. 6). Psychoanalysis is an exemplary psychological version of Western individualistic thought.

The history of psychoanalysis is the story of competing and often opposing visions of why humans suffer and how they might be helped.

Each school of psychoanalysis offers a different account of the former and a different answer to the latter. Since Freud made his ground-breaking discoveries over 100 years ago, psychoanalytic theory and practice have evolved into a continuously expanding framework for the understanding of the mind and the treatment of psychopathology. Five main psychoanalytic viewpoints have been espoused: drive–conflict theory, developmental ego psychology, Sullivanian interpersonal psychoanalysis, British object relations theory, and self psychology. Each of these schools of thought focuses on and illuminates certain crucial aspects of theory and technique.

Greenberg and Mitchell (1983) have suggested that two distinct and incommensurable paradigms underlie these schools: the drive–structure model and the relational model. Classical psychoanalysis and ego psychology are examples of the former, while object relations theory, interpersonal psychoanalysis, and self psychology are exemplars of the latter.

In the drive–structure model, humans are viewed as driven, autonomous, conflicted creatures who are shaped by their historical past and struggle to mediate between endogenously arising, asocial, somatically based, hedonic impulses and the demands of conscience and external reality. Creating a neutral, nonimpinging environment for the disavowed past to emerge, the drive–structure model offers the patient an opportunity to gain a measure of clarity and control over their troubling past. Drive theory was intellectually compatible with the intellectual Weltanschauung of the world in which psychoanalysis arose with its Darwinian flavor, philosophy of science and brain physiology, and neuroanatomy (Mitchell, 1988).

Since the late 1940s, there has been an increasing emphasis in psychoanalysis on relations with others, past and present, actual and imaginary, as an alternative framework for understanding human development as well as the therapeutic process. Within the heterogeneous relational fold, some have emphasized self-experience and organization (Kohut, 1977), some attachment (Bowlby, 1969), and some interpersonal interactions (Sullivan, 1953). From Melanie Klein's (1975) depiction of the human struggle between malevolence and envy, love and gratitude, to Ronald Fairbairn's (1952) account of the human propensity for seeking attachment and relatedness, to Winnicott's (1960) illumination of the human struggles to fashion a life of authenticity and aliveness rather than conformity and deadness, to Kohut's (1977, 1984) emphasis on the development and vicissitudes of self-organization in the matrix of parental responsiveness, the reality and ubiquity of relationships have placed a conceptual strain on the drive model. Many studies comparing psychoanalysis and Buddhism unfortunately equate the former with the drive–structure model and neglect newer views of development, self, and the therapeutic process dictated by relational theories.

Freud (1917) claimed that psychoanalysis was the third great blow to humankind's narcissism. Copernicus demonstrated that the earth was not the center of the universe, Darwin revealed that humans are descendants of animals, and psychoanalysis illuminated the myriad ways that humans are unconscious of vast facets of their thoughts, feelings, fantasies, and conduct and thereby not even masters of their own minds (cf. Freud, 1917, p. 143). The notion of unconsciousness, as Phillips (1994) notes, makes a mockery of the idea of self-mastery. We are shaped by hidden motivations and often behave in ways that are counter to our avowed wishes and conscious intentions. Buddhist teachers and students, no less than analysts and patients, may, for example, engage in self-compromising behaviors or treat others in ways that clash with their conscious ideals. Or, to cite another possible example, our characteristic ways of pursuing freedom may inadvertently lead to self-enslavement, as we place a Buddhist or psychoanalytic doctrine or a Buddhist teacher–psychoanalyst on a pedestal and trust them more than ourselves.

It was Freud's singular genius to detect and articulate the variety of ways that we are self-delusive and opaque to ourselves. Since Freud, we are more aware, for example, of the variety of strategies or defensive processes (Kohut, 1984) that we employ to ward off pain and buttress self-esteem such as denial, repression, projection, displacement, intellectualization, rationalization, and identification with the aggressor. We may, for example, deny and remain silent about intolerance or corruption in the psychological or spiritual communities we inhabit so as not to "rock the boat" with colleagues or friends and risk censure or disapproval. Or, we may rationalize or intellectualize away limitations in the psychotherapeutic or spiritual disciplines we are committed to out of a fear of being independent and free.

Since Freud, we are much more aware of the way that our present is deeply shaped by our past. Because of a ubiquitous psychological phenomenon Freud termed *transference*, we may perceive figures we encounter in the present, including analysts or Buddhist teachers, in a manner that does not befit them, as if they are formative figures from our past. The Buddhist teacher's detachment can be interpreted by his or her students as if, among other things, it is a sign of a negligent parent. The analyst's silence can be experienced as if it is the condemnation of a critical parent.

Not only do we transfer onto others in the present thoughts, feelings, and fantasies we originally experienced with significant others in our past, we may unconsciously *reenact* such forms of relatedness or nonrelatedness (cf. Sandler, 1976). The spiritual seeker who was sexually traumatized as a child may, for example, not only fear a traumatogenic repetition with people in the present, which might include his or her spiritual teacher, he or she might move toward unconsciously recreating such situations.

Psychoanalysis has taught us that analysts, as well as Buddhist teachers, have countertransferences or counterreactions to their patients' and students' transferences. Analysts who were treated as children as if they were invisible might experience a patient's self-centeredness as an intolerable perpetuation of a disturbing form of interpersonal connection. The spiritual teacher who was sexually exploited or starved for emotional closeness or sustenance as a child might respond to a student's interest as a longed-for validation of his or her being, which might result in initiating an intimate relationship and thus enacting and perpetuating his or her earlier history and trauma rather than exploring its meaning and impact on his or her current life.

Another important aspect of psychoanalysis is its recognition of psychic complexity or multidimensionality (Stolorow, Brandchaft, & Atwood, 1987); the fact that any aspect of emotional life may have multiple causes and meanings as well as serve multiple functions (cf. Waelder, 1936). Generosity may hide ambitiousness and self-aggrandizement, and self-denial and self-abasement may also partake of nonattachment in the spiritual sense of the term.

The special conditions of the psychoanalytic situation are designed to promote the optimal unfolding of the patient's unconscious subjective life. Freud's investigative method consisted of a transformative context and a special methodology for investigating and illuminating conscious and unconscious aspects of human subjectivity. The context I am referring to is the self-reflexive dialogue of analyst and analysand and the methodology is the special way of speaking and listening that the analysand and the analyst engage in. The analysand "free associates" or says whatever comes to mind without concern for social propriety or logical coherence. The analyst listens to the analysand with a special quality of heightened attentiveness that Freud (1912) termed "evenly hovering attention."

The analytic situation and the methodological principle of speaking and listening with a minimum of constraints and preconceptions creates an altered state of consciousness for both analyst and analysand that is akin to an imaginative, dreamlike state. This encourages the optimal emergence of the patient's characteristic patterns of seeing and relating to him- or herself and others as well as the analyst's capacity for creative listening. Although this process is literally unpredictable, it often has a similar result.

As the analysand is less concerned with logical consistency, social decorum, self-judgments, pride, and shame, his or her thinking and speech take on a more spontaneous and unfettered form. This opens up the possibility of experiencing previously hidden aspects of self. The patient's discourse is not transparent and lacks a self-evident meaning. The analyst's conception of the analysand is shaped, at least in part, by the analyst's own theoretical models and desires. The Freudian method thus leads not to the truth about who the patient really is but to a variety of ways, including formerly unconscious ones, of conceiving of his or her life. The recurrent unconscious principles and patterns of relating to self and other from the patient's past that shape and delimit him or her, or transference, then appear with greater clarity.

In recommending that the analyst listen afresh to the patient and her associations—and, as it were, dream along with the analysand—the Freudian method can destabilize "fixed" theories including Freudian ones. When the analyst's understandings of the patient's material emerge out of the mutual dialogue between analyst and analysand, a non-authoritarian psychoanalytic climate is promoted and the analyst is encouraged to have a less narcissistic relationship to his or her theories and practices (Rubin, 1995).

Contradictions, gaps, inconsistencies, and displacements in the taken-for-granted narrative that the analysand brings to analysis are more readily recognized when the analyst listens in this way to the analysand's free associations. As alternative conceptions of one's life become possible, one's sense of one's self becomes more complex and less rigid and one-sided (Schafer, 1992). This alters the analysand's sense of her past as well as enriches the possibilities for her present and future. When excessive and inappropriate guilt or shame, for example, are analytically questioned and ultimately mitigated, one may exchange a life lived under an oppressive cloud for an undreamt of sense of freedom. Or, as the unconscious apathy or passivity resulting from disclaimed responsibility for one's life becomes more conscious, a greater sense of personal responsibility and agency may flourish.

Psychoanalysis can foster an enriched sense of "I-ness," an enhanced mode of self-care, and improved kinds of relatedness. In the concluding section of this chapter I shall reflect on what we might learn from our review of psychoanalytic and Buddhist theory and practice.

PSYCHOANALYSIS AND BUDDHISM

"It is just as absurd," claims Hegel (1952), "to imagine that a philosophy can transcend its contemporary world as it is to fancy that an individual can overleap his own age, jump over Rhodes" (p. 11). Psychoanalysis and Buddhism are usually decontextualized from their historical and personal roots and treated as universally valid insights about human life. When they are disconnected from both the specific cultures in which they developed—fifth-century BC India and nineteenth-century Europe—and the particular conditions and difficulties that they were designed to address (e.g. suffering and narcissism and dehumanization and alienation, respectively), it is easier to believe that they offer eternally satisfactory solutions to dilemmas that peoples in different cultures and ages such as our own confront. Unfortunately, this may not always be so.

Let us briefly reflect on certain important historical and personal influences on both traditions so as to more properly contextualize them. This will help us more readily evaluate what they might offer people in late twentieth-century North American and European culture.

Earlier I argued that Buddhism was a kind of Indian Protestantism in the sense of witnessing against and protesting various facets of Hinduism such as authority and tradition, speculation, grace, a personal God, and narcissism. In this section of the chapter I will say a little more about Buddhism's reaction against narcissism and then add one more important factor that Buddhism appears also to have been reacting against: history, particularly the traumatic fact of human finitude and mortality.

Buddhism's critique of egocentricity and advocacy of a life of detachment was certainly salutary. It provided a path out of the jungle of suffering and narcissism that pervaded Buddha's India. Detachment became a strategy for dealing with the dearth of available possibilities for living.

Tolerance and ethics cannot flourish in a world in which egocentricity reigns. A less solipsistic and more compassionate way of being is important in an age like our own that is pervaded by a sense of scarcity, rampant selfishness, and hardheartedness. Buddhism's concern for the other as well as the self can offer an alternative perspective to the politics of blame and scapegoating that permeate contemporary reflections on society.

"The central construct in a theorist's account of human nature and the human condition," as Atwood (1983) notes, "mirrors his [or her] personal solutions to the nuclear crises of his [or her] own life history"

(p. 143). The central construct in Buddha's account of human nature and the human condition is the nonexistence of the self in a world of tremendous suffering. The central conflict that seems to have thematized his young adult life was the horror and fear occasioned when he first encountered as a sheltered young man the human misery and mortality that are a part of human history. He found old age, disease, and death terrifying and profoundly disturbing. It filled him with disgust and dread. He felt life was "oppressive and stifling," enslaving and debilitating. Life was, in a word, deadly. He responded to this terrifying specter by immediately fleeing from the world and pursuing a renunciate life. He then formulated a death-defying philosophy–psychology of self-denial and self-immortalization: He both denied the reality of personhood—" 'Misery only doth exist, none miserable' " (Buddhaghosa, 1976)—and affirmed the possibility of a "deathless" realm beyond history, human suffering, and mortality: "Vigilance is the abode of eternal life," Buddha says in the *Dhammapada*, "thoughtlessness is the abode of death. Those who are vigilant (who are given to reflection) do not die" (1950, p. 66).

But of course this is quixotic because there is no deathlessness in a human life. No one eludes death. Buddhism's strategy for coping with this intractable fact is to depersonalize mind and human history by viewing them as, in Althusser's words, a process without a subject or self, thereby denying the reality of human existence. By eliminating human existence, which obviously disengages subjects from affective life and the world, Buddhism attempts to avoid and ward off human finitude and misery.

Whether or not the revolutionary strategy Buddha fashioned for addressing the historicity of human existence did or did not partake of a defensive escape from excruciatingly painful realities of human existence that might be worthy of further investigation with his meditative method of radical self-inquiry, what does seem clear is that it had within it some potential liabilities. With its emphasis on selfhood's insubstantiality, Buddhism promotes self-nullification, which can preclude questions of human agency and inhibit political engagement. For if there is no subject, then there is no one who is exploited or alienated and no oppression to challenge or contest. This will be explored in more detail in Chapter 3.

The emphasis in psychoanalysis on the development of the psyche of the "psychological" subject offered psychological grounding and comfort for the psychologically dehumanized and alienated early twentieth-century citizen. But its individualistically oriented psychological theory of persons can promote an excessively self-centered view of self

(Rubin, 1995) and an incomplete view of relationships and morality. In Chapter 10, I shall attempt to demonstrate that this may eclipse aspects of self-transcendence and spirituality, that which goes beyond the autonomous, separate self, as well as foster a sense of alienation, disconnection, and isolation. A world that excludes spirituality seems impoverished. The increasing nihilism, alienation, and disconnection in our world may be related, at least in part, to our neglect of the spiritual.

The conceptions of subjectivity that arose to deal with the narcissism and suffering of Buddha's India and the threats to identity in Victorian Europe are particular and partial. In overemphasizing, respectively, detachment from affective life and a reified, solidified, egoistic–possessive individualism, Buddhism and psychoanalysis each necessarily eclipse certain of subjectivities features, and possibilities and complicate coping with certain facets of late twentieth-century life in the West.

(UN/POST?) MODERN TIMES

We live in different and difficult times. Certain of the familiar coordinates by which previous ages mapped themselves have been challenged in recent years. In our world there are no absolute foundations for knowledge; there is no totalizing logic of the social world, by which I mean that it is impossible to describe the social structure from the vantage point of a "single, or universal point of view" (Laclau, 1988; Mouffe, 1988); there is a despiritualization (Berman, 1981) and functional stratification (Luhmann, 1986) of the social landscape, with individuals experiencing multiple "subject positions" (Mouffe, 1988) or a plethora of kinds of roles and self-states; and there is a sense of profound rootlessness, disconnection, and ontological "homelessness" (Berger, Berger, & Kellner, 1973). Let me elaborate.

An important consequence of modernism is a sense of what Max Weber termed a "disenchantment" of the world. In the premodern world it was widely assumed that the universe was an intrinsically meaningful and hierarchically ordered whole, with every entity and form of being, including animate and inanimate objects, animals, humans, and God, having a preordained status, significance, and function. One's responsibility was to ascertain and live in accordance with this inherited social and moral framework, which was constitutive of the person's identity and lent a coherence and direction to human lives.

The ascendancy of modernism involved a discrediting of this conception of the world. But the increased individual freedom resulting from decreasing allegiance to this inherited sacred order was purchased at the cost of eliminating the comforting, meaning-giving moral and cosmic horizons that had guided human choices and actions. Without the social and cosmic framework of reference that had guided people in the premodern world, the conduct of modern citizens became anormative, devoid of a shared standard of values and conduct. Instead, human conduct often became guided by individualistic and essentially self-centered needs and attitudes. This led, as I suggested in the introduction, to both "flattened" and "narrowed" lives of "self-absorption" (cf. Taylor, 1991, p. 4) and a sense of personal alienation.

We inhabit a world that is, according to Luhmann (1986), "functionally differentiated" rather than "stratified" (p. 318). In stratified societies, the individual is ordinarily placed "in only one subsystem" based on "social status (condition, qualité, état) [which] was the stable characteristic of an individual's personality." In a society like ours, which is "differentiated with regard to politics, economy, intimate relationships, religion, sciences, and education," this is no longer possible. Nobody lives "in only one of these systems" (Luhmann, 1986, p. 318). In fact, we often live in many.

A single person thus inhabits a "multiplicity of subject-positions" (Mouffe, 1988, p. 34). We are always "multiple and contradictory subjects," who inhabit "a diversity of communities," depending on such things as "the social relations in which we participate and the subject-positions they define." We are thus "constructed by a variety of discourses and precariously and temporarily sutured at the intersection of those subject-positions" (Mouffe, 1988, p. 44). A person is thus, in Pynchon's (1973) evocative phrase, a "crossroads, a living intersection" (p. 625) of multiple psychological, sociocultural, and historical influences, intersecting in complex and sometimes conflicting ways.

There is some evidence that a configuration that Cushman (1990) terms the "empty self" is a subject position that predominates in the post-World War II era in the United States. By "empty self," Cushman means a self that experiences a significant "absence of community, tradition and shared meaning.... It is empty in part because of the loss of family, community and tradition.... It experiences these social absences and their consequences 'interiorly' as a lack of personal conviction and worth, and it embodies the absences as a chronic undifferentiated emotional hunger.... It is a self that seeks the experience of being continually filled up by consuming goods, to acquire and consume as an unconscious way of compensating for what has been lost: It is empty" (p. 600).

These unprecedented features in our postmodern world necessitate more expanded conceptions of subjectivity that neither psychoanalysis' focus on the self-centered individual nor Buddhism's emphasis on the agentless subject can sanction, let alone imagine. This will be explored in Chapter 3. In the next chapter, I shall examine how other thinkers have viewed psychoanalysis and Buddhism and then present the perspective I shall be utilizing throughout this book.

REFERENCES

Atwood, G. (1983). The pursuit of being in the life and thought of Jean-Paul Sartre. *Psychoanalytic Review, 70*(2), 143–162.

Basham, A.L. (1988). Jainism & Buddhism. In A. Embree (Ed.), *Sources of Indian tradition, Vol. I: From the beginning to 1800* (pp. 41–199). New York: Columbia University Press.

Baumeister, R. (1987). How the self became a problem: A psychological review of historical research, *Journal of Personality and Social Psychology, 52,* 163–176.

Berger, P., Berger, B., & Kellner, H. (1973). *The homeless mind.* New York: Random House.

Berman, M. (1981). *The reenchantment of the world.* New York: Bantam Books.

Bowlby, J. (1969). *Attachment.* New York: Basic Books.

Buddha, G. (1950). *Dhammapada.* London: Oxford University Press.

Buddhaghosa, B. (1976). *The path of purification.* Berkeley, CA: Shambhala.

Burtt, E. (1955). *The teachings of the compassionate Buddha.* New York: New American Library.

Carrithers, M. (1983). *The Buddha.* New York: Oxford University Press.

Carrithers, M. (1985). An alternative social history of the self. In M. Carrithers, S. Collins, & S. Lukes (Eds.), *The category of the person* (pp. 234–256). Cambridge, England: Cambridge University Press.

Cushman, P. (1990). Why the self is empty: Toward a historically situated psychology. *American Psychologist, 45,* 599–611.

DeBary, W. (1972). *The Buddhist tradition in India, China, and Japan.* New York: Random House.

Fairbairn, R. (1952). *An object-relations theory of the personality.* New York: Basic Books.

Freud, S. (1985). Studies on hysteria. In J. Strachey (Ed. & Trans.), *The standard edition of the complete psychological works of Sigmund Freud,* Vol. 2 (pp. 255–305). London: Hogarth Press.

Freud, S. (1912). Recommendations to physicians practicing psycho-analysis. In J. Strachey (Ed. & Trans.), *The standard edition of the complete psychological works of Sigmund Freud,* Vol. 12 (pp. 111–120). London: Hogarth Press.

Freud, S. (1917). A difficulty in the path of psycho-analysis. In J. Strachey (Ed. & Trans.), *The standard edition of the complete psychological works of Sigmund Freud,* Vol. 17 (pp. 135–144). London: Hogarth Press.

Goleman, D. (1977). *The varieties of the meditative experience.* New York: E.P. Dutton.

Greenberg, J., & Mitchell, S. (1983). *Object relations in psychoanalytic theory.* Cambridge, MA: Harvard University Press.

Hegel, G. (1952). *Philosophy of right* (T.M. Knox, Trans.). Oxford, England: Clarendon Press.

Klein, M. (1975). *Love, guilt & reparation & other works 1921–1945.* London: Hogarth Press.

Kohut, H. (1977). *The restoration of the self.* New York: International Universities Press.

Kohut, H. (1984). *How does analysis cure?* Chicago: University of Chicago Press.

Kornfield, J. (1977). *Living Buddhist masters.* Santa Cruz, CA: Unity Press.

Kovel, J. (1991). *History and spirit: An inquiry into the philosophy of liberation.* Boston: Beacon Press.

Laclau, E. (1988). Politics and the limits of modernity. In A. Ross (Ed.), *Universal abandon?: The politics of postmodernism* (pp. 63–82). Minneapolis: University of Minnesota Press.

Luhmann, N. (1986). The individuality of the individual: Historical meanings and contemporary problems. In T. Heller, M. Sosna, & D. Wellbery (Eds.), *Reconstructing individualism: Autonomy, individuality, and the self in Western thought* (pp. 313–324). Stanford, CA: Stanford University Press.

McNeill, W. (1963). *The rise of the West: A history of the human community.* Chicago: University of Chicago Press.

Mitchell, S. (1988). *Relational concepts in psychoanalysis: An integration.* Cambridge, MA: Harvard University Press.

Mouffe, C. (1988). Radical democracy: Modern or postmodern? In A. Ross (Ed.), *Universal abandon: The politics of postmodernism* (pp. 31–45). Minneapolis: University of Minnesota Press.

Phillips, A. (1994). *On flirtation: Psychoanalytic essays on the uncommitted life.* Cambridge, MA: Harvard University Press.

Pynchon, T. (1973). *Gravity's rainbow.* New York: Viking Press.

Rieff, P. (1963). Introduction. In *Freud: Therapy and technique.* New York: Macmillan.

Roccasalvo, J. (1982). The terminology of the soul (atta): A psychiatric recasting. *Journal of Religion and Health, 21*(3), 206–218.

Roland, A. (1995). *How universal is the psychoanalytic self?* Unpublished manuscript.

Ross, N. (1966). *Three ways of Asian wisdom: Hinduism, Buddhism and Zen and their significance for the West.* New York: Simon and Schuster.

Rubin, J. B. (submitted). *The blindness of the seeing I: Perils and possibilities in psychoanalysis.* New York: New York University Press.

Sandler, J. (1976). Countertransference and role-responsiveness. *International Journal of Psycho-Analysis, 3,* 43–47.

Schafer, R. (1992). *Retelling a life: Narration and dialogue in psychoanalysis.* New York: Basic Books.

Smith, H. (1986). *The religions of man.* New York: Harper & Row.

Stolorow, R., Brandchaft, B., & Atwood, G. (1987). *Psychoanalytic treatment: An intersubjective approach.* Hillsdale, NJ: The Analytic Press.

Stryk, L. (Ed.). (1968). *World of the Buddha.* New York: Grove Weidenfeld.

Sullivan, H.S. (1953). *The interpersonal theory of psychiatry.* New York: Norton.

Taylor, C. (1991). *The ethics of authenticity.* Cambridge, MA: Harvard University Press.

Waelder, R. (1936). The principle of multiple function. *Psychoanalytic Quarterly, 35,* 45–62.

Warren, H.C. (1977). *Buddhism in translations.* New York: Atheneum.

Winnicott, D.W. (1965). Ego distortion in terms of true and false self. In D.W. Winnicott, *The maturational processes and the facilitating environment* (pp. 140–152). New York: International Universities Press.

2

Beyond Eurocentrism and Orientocentrism

Religion has played a central role in human history. Definitions of reality, wars, and visions of moral excellence have been part of its complex legacy. Religion has been involved in a variety of human activities including prohibiting murder, extolling poverty and renunciation, distributing power, and regulating procreation (Pruyser, 1973). A history of religion might include accounts of compassion and persecution, wisdom and fanaticism. In our world, religion is Janus-faced, a progenitor of evil, spawning psychotic cults and unconscionable violence and a midwife to psychological transformation, facilitating growth and compassion.

The relationship between psychoanalysis and religion has been fraught with conflict and misunderstanding. Religious issues have been avoided in much clinical diagnosis and treatment. Paul Pruyser (1971) notes that case records are "conspicuously devoid of articulate reference to religion" (p. 272). There is a "conspiracy of silence" about religion in both diagnostic interviewing and in psychotherapy (p. 272).

This selective inattention to religion on the part of clinicians becomes striking when it is juxtaposed with a 1984 Gallop Poll finding that 99% of the American public reports "a belief in God and a belief in prayer" and with the Group for Advancement of Psychiatry's (1968) report that "manifest references to religion occur in about one third of all psychoanalytic sessions" (p. 54).

Psychoanalytic clinicians, for the most part, seem highly uncomfortable with religion. In the words of William Meissner (1984), a psychoanalyst and a Jesuit, psychoanalysts "tend to regard religious thinking and convictions as suspect, even to hold them in contempt at times" (p. 5). According to Meissner there is a "latent persuasion, not often expressed or even articulated ... that religious ideas are inherently neurotic, self-deceptive and illusive" (p. 5).

Psychoanalysis has historically adopted a biased and restrictive perspective on religion, asserting, without always clinically demonstrating, that it is pathological and maladaptive. This antireligious bias within psychoanalysis began with Freud's writings on religion. Critiques of religion in Western Europe's Enlightenment antedated Freud, but it was his contribution to expose the parallels between an individual's personal history and the later configuration of his or her neurotic religious beliefs and practices (Lovinger, 1989).

Freud's interest in religion was evident throughout his work. Several recurrent themes dominate his disparate reflections on religion. Freud saw himself as a destroyer of illusions. Writing to the poet Romain Rolland, a student of the renowned Indian saint Ramakrishna, he says, "a great part of my life's work ... has been spent [trying to] destroy illusions of my own and those of mankind" (Meng & Freud, 1963, p. 341).

Freud called religion many things—a universal obsessional neurosis (1927, p. 43); a childhood neurosis (1927, p. 53); a form of masochism (1930); a reaction formation against unacceptable impulses (1927); a "delusion" (1927, p. 31)—but above all it was, for him, an "illusion" (1927, p. 30), an unrealistic belief that contradicts experience and reason. Illusion, for Freud, was not an error but a "wish-fulfillment" (1927, pp. 30–31). While he points out that illusion is not necessarily "false" and that "the truth-value" of religious doctrines does not lie within the scope of psychoanalytic inquiry (1927, p. 33), he nonetheless proceeds to condemn it as comparable to a "childhood neurosis" (1927, p. 53).

Two childhood wishes or psychological needs seemed to lead people to construct religious beliefs: the necessity of coming to terms with the complicated emotions of a child's relation to his or her father and the child's sense of helplessness in the face of the danger of the inner and outer worlds.

Helplessness arouses the need for protection. Religious ideas, according to Freud, are born of the need to make tolerable the human sense of helplessness. They are designed to offer compensation, consolation, and protection for our existential vulnerability. Religion "allays our fear of the dangers of life" (1927, p. 33).

The sociology of religion echoes Freud's view of religion. Historical evolution, according to Auguste Comte's "law of three," proceeds from myth–religion to metaphysics to rational science. Here, religion is viewed as a "primitive consolation for a primitive mentality" (Wilber, 1983, p. 8). The sequence of magic to myth to rationality is not fictitious.

Horrors such as Jonestown or the assassination of Itzhak Rabin demonstrate that religious movements can be expressions of infantile, pathological beliefs and thinking. The problem is that not all religious involvements are childish and immature and the meaning, function, and purpose of one's involvement in religion are effaced by such monolithic and reductionistic interpretations.

Freud believed that all sorts of "crimes" (1927, p. 32) were committed in the name of religion: "when questions of religion are concerned, people are guilty of every possible sort of dishonesty and intellectual misdemeanor" (1927, p. 32). He hoped that humankind would surmount this "neurotic phase" and attain an "education to reality" (1927, p. 49). Like the Roman poet Lucretius, he wished to awaken humankind from the enchantment in which the priests held it captive (Gay, 1987). But Freud believed, to borrow a metaphor he used in contrasting psychotherapy with psychoanalysis, that only a select few are capable of replacing the superficial salve of religious illusions with the "pure gold" of psychoanalysis.

Freud explicitly acknowledged that his views on religion were his own and "form no part of analytic theory ... there are certainly many excellent analysts who do not share them" (Meng & Freud, 1963, p. 117). His knowledge of religion seemed confined to the Old and New Testaments and the Greek, Roman, and Egyptian religions of antiquity. His theoretical writings on religion focused on Judaism and Christianity. His colleague and biographer Ernest Jones suggests that Freud did not appear to have "extended these studies to the religions of India and China" (Jones, 1957, p. 351). According to Jones, Freud admitted that his study of religious belief was limited to that of the common man, and that he regretted having ignored the rarer and more profound type of religious emotion as experienced by mystics and saints (Jones, 1957, p. 360). In *Civilization and Its Discontents*, Freud (1930) states that in *The Future of an Illusion* (1927) he was "concerned much less with the deepest sources of religious feeling than with what the ordinary man understands by his religion—with the system of doctrines and promises which on one hand explains to him the riddles of this world with an enviable completeness, and on the other, assures him that a careful Providence will watch over his life and will compensate him in a future existence for any frustrations he suffers here" (1930, p. 74).

Freud did not explore mature spirituality in general or Asian religion in particular. He buried religion alive.

Some aspects of Freud's work on religion such as his anthropological speculations about and patricentric explanations of the formation of God representations have met with skepticism and dismissal. For the most part, his viewpoints on religion have, in the words of Saffady (1976), "held full sway in the psychoanalytic literature" (p. 292). Subsequent psychoanalysts have "tended to disagree with him in details more than essentials" (p. 292). Freud's own skepticism regarding religious beliefs and practices have been mirrored by the vast majority of subsequent psychoanalysts who have put more nails in religion's coffin. The subsequent psychoanalytic literature has, for the most part, continued his predilection for pathologizing and dismissing religious experience.

Yet a small number of psychoanalysts (e.g. Pfister, 1948; Silberer, 1917; Jung, 1958a; Menninger, 1942; Horney, 1945, 1987; Kelman, 1960; Fromm, Suzuki, & DeMartino, 1960; Milner, 1973; Rizzuto, 1979; Meissner, 1984; Kohut, 1985; Rubin, 1985, 1991, 1993; Winnicott, 1986; Roland, 1988; Finn, 1992; Suler, 1993) have embraced psychoanalytic theory and technique while attempting to nonreductively explore religious phenomena. Bion (1970) was one of the only analysts other than Jung to introduce mysticism into psychoanalysis. He was also arguably the most mystical psychoanalyst (cf. Roland, 1988, p. 293). But since his valuable contribution to psychoanalysis and mysticism lies in a different realm—psychoanalytic epistemology and listening and mysticism and psychoanalytic groups—I will discuss him in Chapter 6 rather than in this chapter.

Only a few analysts have actually attempted to integrate psychoanalytic thought with religion in a public as opposed to a private manner. A summary review of these and related attempts to think about Western psychotherapeutic and Eastern contemplative disciplines in tandem may help place the current encounter of psychotherapy and psychoanalysis and Buddhism into perspective. Exploring nonanalytic thinkers such as Alan Watts, the Italian psychosynthesist Roberto Assagioli, humanistic psychologist Abraham Maslow, the existentialist psychiatrist Medard Boss, and transpersonal theorists Jack Engler and Ken Wilber will also aid us in this task.

The Swiss psychoanalyst and pastor, Oskar Pfister, a close personal friend of Freud, was one of the first respondents to Freud's attacks on religion. Pfister (1948) concurred with Freud's observation that unconscious wishes may color the development of religion but believed this was insufficient to explain all of religion. In fact, he maintained that these motivations are more characteristic of less evolved forms of religion. According to Pfister, mature religious practice involved the exact opposite of the infantilism and self-centeredness so prevalent in its less evolved manifestations. Pfister believed that Freud ignored religion's most essential features and its noblest reflections.

Among psychoanalysts and personality theorists, Carl Jung was most knowledgeable about Eastern psychologies. He had a long-standing interest in Eastern thought and a great respect for certain Oriental teachings. Eastern thought had a significant and beneficial impact on his life and work.

Jung's concern with Eastern religion and philosophy was evident in his theoretical work, *Wandlungen und Symbole der Libido* (1912; revised in 1952 as *Symbols of Transformation*), in which he utilized symbolic material from the religious traditions of India. His first separate work in the genre was his commentary on *The Secret of the Golden Flower* (1929), which is both a Taoist text concerned with Chinese yoga and an alchemical treatise. Jung wrote forewords to and commentaries on traditional Eastern texts: a psychological commentary on *The Tibetan Book of the Dead* and the *Tibetan Book of the Great Liberation*; a foreword to the I Ching; an evaluation of the discourses of the Buddha; and an essay, "The Holy Men of India," written to introduce the Indologist Heinrich Zimmer's collection of the teachings of the Hindu saint Ramana Maharshi. In addition, Jung wrote forwards to books on Eastern thought by contemporaries, including Zen scholar D.T. Suzuki's (1974) *An Introduction to Zen Buddhism*.

For Jung, civilizational health, like individual health, was based in large part on balancing opposing aspects of life. Psychic one-sidedness, in Jung's view, may produce short-term gains of specialization, but in the long run is emotionally detrimental. "One-sidedness, though it lends momentum, is a mark of barbarism" (Jung, 1929, p. 9). The modern West, according to Jung, is one-sided. After having placed emphasis on the spiritual and intuitive during the Middle Ages, the intellect has assumed a position of dominance in the modern world. Other aspects of human experience have been neglected and excluded. For Jung, we are too one-sided with our overemphasis on intellect and neglect of the intuitive. From the study of Eastern thought, Western culture could rediscover or resensitize itself to dimensions of life that have been neglected and excluded, such as intuition, feelings and the inner life. While Jung admired the Eastern principle of inclusiveness and balance, he felt that the East had overstressed the intuitive at the expense of its sensitivity to science and technology.

Jung warned of the dangers for a Westerner of involvement in Eastern traditions. He felt that Eastern practices were unsuited to the Western mind. In "Yoga and the West," he says, "Study yoga—you will learn an infinite amount from it—but do not try to apply it, for we Europeans are not so constituted that we apply these methods correctly, just like that" (1958b, p. 534). Eastern practices would strengthen a patient's will and consciousness and thus perpetuate and further intensify their split from the unconscious. This would further aggravate the already chronic Western overdevelopment of the will and the conscious aspect of the personality. Another danger of uncritical adoption of Eastern practices is that many Westerners would become entranced by the exotic teachings of the East and thus avoid their own problems:

> People will do anything, no matter how absurd, in order to avoid facing their own souls. They will practice yoga and all its exercises ... learn theosophy by heart, or mechanically repeat mystic texts from the literature of the world— all because they cannot get on with themselves and have not the slightest faith that anything useful could ever come out of their own souls. (Jung, 1968, pp. 99–101)

Rather than importing foreign methods, we should, according to Jung, find our own path: "Instead of learning the spiritual techniques of the East by heart and imitating them ... it would be far more to the point to ... build on our ground with our own methods" (1958a, p. 483).

Silberer detected a constructive dimension in religious experience. He distinguished between true mysticism and false mysticism. The latter is characterized by an "extension of personality," the former by a "shrinking" (Milner, 1973, p. 265).

Karl Menninger (1942) also took exception to psychoanalysis' exclusive emphasis on the neurotic determinants of religious belief. Agreeing with Freud's claim that transference, human infantilism, and egocentricity play an important role in the power of religious belief, he also stresses religion's adaptive aspects such as its capacity to "control and direct aggression" and "foster life by inspiring love" (Menninger, 1942, p. 191).

Apart from Jung, the Eastern psychologies have made their greatest impact in Western therapy through their influence on such theorists and clinicians as psychoanalysts Karen Horney and Harold Kelman, the psychological theorist Alan Watts, the humanistic, existential psychoanalyst Erich Fromm, the existential psychiatrist Medard Boss, the Italian psychosynthesist Roberto Assagioli, the humanistic psychologist Abraham Maslow, and the transpersonal theorist Ken Wilber.

The esteemed analyst Karen Horney became increasingly concerned with ultimate and spiritual questions and investigated Zen Buddhism in the last years of her life. Horney (1945) first drew on Zen in *Our Inner Conflict*. Quoting from D.T. Suzuki, a Japanese scholar, writer, and esteemed interpreter of Zen to the West, she discussed Zen writings on sincerity and wholeheartedness. The goal of therapy, for Horney, was "wholeheartedness: to be without pretense, to be emotionally sincere, to be able to put the whole of oneself into one's feelings, one's work, one's beliefs" (1945, p. 242). In *Our Inner Conflicts*, she referred to a "most interesting" paper by William James, "The Energies of Man," which discusses a number of examples of how spiritual training helped overcome illness. Horney's concern with ultimate and spiritual questions and interest in non-Western thought grew toward the end of her life. She read from Aldous Huxley's (1944) *The Perennial Philosophy*, a compendium of major spiritual traditions including Islamic, Christian, and Buddhist writings, almost every night the last 4 years of her life (Quinn, 1987). Shortly before her death she traveled to Japan with D.T. Suzuki to explore how Zen might reaffirm or complement her own theories.

In a talk at Jikei-Kai Medical School in Tokyo, Horney suggested that there were a number of similarities between her ideas and those of Shomo Morita, the founder of Morita therapy, a treatment used in Japan that borrowed heavily from Zen. According to Morita, patients' problems resulted from their being captives of their own "subjectivity" and egocentricity. The cure drew on Zen and encouraged acceptance of things as they are.

In her posthumously published *Final Lectures*, Horney (1987) discusses how Buddhism can enrich psychoanalysis. For Horney, as for the vast majority of analysts since Freud, the analyst's "attentiveness to the patient" is the basic prerequisite for doing sound analytic work. Attentiveness, according to Horney, is a "rare attainment" (1987, p. 35), a "faculty for which the Orientals have a much deeper feeling than we do ... [and] a much better training" (p. 18). Buddhism can train what Horney terms "wholehearted attentiveness" or absorption in what one is doing. Training of this capacity will be discussed in more detail in Chapter 6. I agree with a recent biographer's reflections that "Had there been more time, there is no telling how Horney might have altered her views as the result of her new interest in Zen" (Quinn, 1987, p. 415).

Harold Kelman, a past president of the American Academy of Psychoanalysis, was a close associate of Karen Horney and served for years as editor of the *American Journal of Psychoanalysis*. Like Horney, Kelman felt an affinity with Eastern thought. In "Psychoanalytic Thought and Eastern Wisdom" (1960), he suggested that psychoanalysis carried to its logical conclusion is Eastern in technique but not in theory. According to Kelman, the theories underlying psychoanalysis and Eastern thought are different, but Eastern techniques can enrich psychoanalytic practice. The way patients and analysts listen is, according to Kelman, very Eastern. Patients are told by their analysts to say whatever comes to mind without judgment or censorship, which cultivates a meditative state of mind.

Alan Watts, though not a psychological theorist or clinician, played a seminal role in bringing Eastern teachings to the awareness of Western psychology in a series of books, most notably *Psychotherapy East and West* (1961), and as a guest lecturer at numerous psychiatric hospitals and medical schools. Watts maintained that the Eastern "ways of liberation," like Western psychotherapy, are focused on altering peoples' feelings about themselves and their relations to others and to nature. Western therapies tend to deal with disturbed people, while Eastern approaches address socially adjusted people. Nevertheless, Watts felt that the goals of the ways of liberation were compatible with the aims of several therapeutic theorists, especially Jung's individuation, Allport's functional autonomy, Adler's creative selfhood, and Maslow's self-actualization (cf. Goleman, 1988).

Medard Boss, a Swiss existential psychiatrist, also greatly valued Eastern thought. His perspective derived from his exposure to the Indian holy men he met when he lectured on psychiatry in India. He came away from these encounters deeply impressed with the sages he met:

> And yet there were the exalted figures of the sages and holy men themselves, each one of them a living example of the possibility of human growth and maturity and of the attainment of an imperturbable inner peace, a joyous freedom from guilt, and a purified, selfless goodness and calmness.... No matter how carefully I observe the waking lives of the holy men, no matter how ready they were to tell me of their dreams, I could not detect in the best of them a trace of a selfish action or any kind of repressed or consciously concealed shadow life. (Welwood, 1979, p. 186)

These meetings convinced Boss that when compared to the behavior of these Eastern sages, the methods and goals of Western psychotherapy were quite limited. In *A Psychiatrist Discovers India*, Boss (1965) suggests that "even the best Western training analysis is not more than an introductory course" compared with the techniques of Eastern systems.

Erich Fromm, an existentially oriented psychoanalyst, had a long-standing history of contact with Eastern teachers. In *Zen Buddhism and Psychoanalysis*, coauthored with D.T. Suzuki and Richard DeMartino, a professor of religion, Fromm (1970) focused on how psychoanalysis can be useful to spiritual practitioners and how meditation can lead beyond the limits of therapy. Psychoanalysts can help meditators avoid "the danger of a false enlightenment ... based on psychotic or hysterical phenomena, or on a self-induced state of trance. Analytic clarification might help ... [meditators] to avoid illusions" (1970, pp. 140–141). Meditation practice can extend the psychoanalytic vision of optimal psychological health. Well-being is a difficult achievement that may, according to Fromm, go beyond the aims of psychoanalysis. Meditation focuses not on the removal of symptoms, an "absence of illness," but the "presence of well-being" (1970, p. 86): being fully born, overcoming narrow views of the self, being completely aware of and responsive to the world. This meditative vision of health, according to Fromm, extends and enriches the aims of psychoanalysis.

The Italian psychiatrist Roberto Assagioli shared a similar viewpoint. Assagioli's (1971) "psychosynthesis" attempted to incorporate perspectives gleaned from both Eastern and Western traditions. He advocated an amalgam of therapeutic methods that address in ascending order a patient's physical disorders, psychological disturbances, and spiritual needs.

Abraham Maslow's (1971) *The Farther Reaches of Human Nature* indirectly carried Alan Watts' work a step further. In "Theory Z" he posits a vision of healthiness that resembles the ideal types of meditators in Eastern psychologies. Although no specific Eastern writings are cited, Eastern concepts are sprinkled throughout his discussion.

In "Some Notes on Psychoanalytic Ideas about Mysticism," psychoanalyst Marion Milner (1973) describes other constructive dimensions of religious experience. Although cognizant of the way religion or what she terms "mysticism" can lead to "mental self-blinding, a dangerous denial of unpleasant truths, both in ourselves and the world" (p. 262), Milner also acknowledges the potentially "recuperative" and enriching dimensions of religion. She describes, for example, sitting in a garden at a residential art school, wishing to paint but unable to find a subject. In order to deal with the tension she had "started a deep breathing exercise and had been astonished to find that the world around me immediately became quite different and by now, exceedingly paintable ... that turning one's attention inward, not to awareness of one's big toe but to the inner sensations of breathing ... [had] a marked effect on the appearance and significance of the world" (pp. 260–261).

In *The Birth of the Living God*, a ground-breaking empirical work on the origin, development, and use of God representations during the course of human life, psychoanalyst and Roman Catholic Ana-Marie Rizzuto (1979) has illuminated the complexity of religion and God representations. In contrast to Freud, who maintained that belief in God is based on a child's creation of a surrogate father figure to make tolerable a sense of helplessness in dealing with the universe, Rizzuto suggests that one's God representation derives from various sources and is a crucial aspect of one's view of self, others, and the world.

In *Psychoanalysis and Religious Experience*, William Meissner (1984), a Jesuit and Boston training analyst, attempts to facilitate a meaningful dialogue between psychoanalysis and religion. Although he does not deal with Asian religion, his remarks apply, with the appropriate changes, to the topics at hand. Meissner contends that psycho-analysis has traditionally adopted a biased and reductionistic perspective in exploring religious experience. For Meissner, Freud's view of religion provided a distorted and fragmentary picture that "precludes the acceptance of a more mature viewpoint" (Cohen, 1986, p. 46).

Meissner, like Menninger, acknowledges that religion can embody regressive components, while also stressing that it can facilitate an amplified and deeper quality of life. He views it in a more healthy light and hopes to renew an authentic dialogue between psychoanalysis and religion.

Heinz Kohut never produced a systematic and comprehensive work on religion, but references to religion are scattered throughout his writings. In a posthumously published interview, "Religion, Ethics and Values," Kohut maintained that religion was a "complex phenomenon" (1985, p. 261) with various meanings and functions. Religion, in Kohut's view, was "poor science" (1985, p. 261): sometimes a "crude mythology on the level of a fairy tale ... in other words, primitive science" (p. 247), but he also felt that psychoanalysis had "underestimated" the "supportive aspect of religion" (p. 261), particularly its "civilizing influence." He maintained that religion could reduce "unmodified narcissism," offer constructive ideals, and curb destructive tendencies.

In "Meditation and Psychoanalytic Listening" (Rubin, 1985), I propose one way that psychoanalysis and Buddhism can be integrated. Freud defined the optimal state of mind for analysts to listen to patients as "evenly hovering attention," but he did not discuss, in a positive sense, how to develop it. He focused on what to avoid (e.g., censorship and prior expectations), not what to do, that is how to cultivate "evenly hovering attention." Buddhist insight meditation cultivates exactly this state of mind. Therefore, it can enrich psychoanalytic listening.

In his posthumously published *Home Is Where We Start From*, Winnicott (1986) asserts that participation in religion is an aspect of a healthy life. Healthy people, according to Winnicott, live in three worlds: (1) "life in the world, with interpersonal relationships;" (2) the life of the "personal (sometimes called inner psychical reality)," and (3) "the area of cultural experience ... including the arts, the myths of history, the slow march of philosophical thought and the mysteries of mathematics, and of group management and of religion" (1986, pp. 35–36). In Winnicott's view, religion, being part of the capacity to have cultural experience, is an essential aspect of health and healthy living.

Alan Roland's (1988) psychoanalytic investigation of the self-experience of Indian and Japanese patients, some of whom are involved in non-Western religions such as Hinduism, represents one of the few sympathetic approaches to non-Western religion in the psychoanalytic literature. Roland aptly notes that with rare exceptions psychoanalysts have viewed religion in terms of "compensations and psychopathology" (1988, p. 59). He recommends that psychoanalysts respect the unique meaning and purpose that religion serves in people's lives. Studying patients who are actively involved in spiritual disciplines can reduce psychoanalytic ethnocentrism and expand psychoanalytic conceptions of subjectivity (Roland, 1988; Rubin, 1993).

The attempt to redress psychoanalytic reductionism and establish a meaningful dialogue between psychoanalysis and religion on the part of the psychoanalysts that I have discussed has provided a crucial service to psychoanalysis. These efforts, however, have rarely been applied to Asian religion.

Traditionally, non-Western psychologies in general and Asian psychology in particular have rarely been discussed in Western psychology. Asian psychology has been virtually ignored in the psychoanalytic literature. Psychoanalytic explorations of religion have almost universally focused on non-Asian religions. The few exceptions (e.g., Alexander, 1931; Masson & Masson, 1978) present distorted and reductionistic accounts of it, falsely equating it with mysticism (Masson & Masson, 1978), regression, and pathology (Alexander, 1931). In his blanket dismissal of Buddhism as a training in an artificial catatonia, psychoanalyst Franz Alexander (1931) illustrates the Eurocentrism that has plagued psychoanalysis.

The more sympathetic non-Eurocentric work of Jung (1958a); Horney (1945, 1987), Kelman (1960), Fromm et al. (1960), Roland (1988), Rubin (1985, 1991, 1992, 1993), and Suler (1993) are the exceptions that demonstrate the rule. Although they have pointed to various aspects of Buddhism's salutary dimensions, including its ability to sensitize us to the inner life (Jung, 1958a), enrich psychoanalytic listening (Rubin, 1985), improve affect demarcation and tolerance (Rubin, 1992), promote "well-being"—being fully awake and alive (Fromm, 1960)—and expand psychoanalytic conceptions of subjectivity (Roland, 1988; Rubin, 1993; Suler, 1993),—they tend, with the exception of the work of Roland (1988), Rubin (1991, 1992, 1993), and Engler (1984), to neglect clinical issues and case material. There are thus few extant precedents for integrating psychoanalysis and Buddhism.

The two most compelling attempts to integrate Asian and Western psychology are Jack Engler's (1984) "developmental" model and transpersonal theorist Ken Wilber's (1977; 1978a,b; 1986) "spectrum psychology." Both thinkers exhibit an exemplary mastery of both traditional psychological theory and spiritual disciplines as well as an integration of theory and practice.

Engler attempts to integrate conventional psychotherapeutic and contemplative spiritual disciplines by seeing them as complementary facets of a developmental continuum, with the former representing "lower" stages of development and the latter representing "higher" stages. Developing a strong, cohesive sense of self is the "precondition" of the contemplative task of disidentifying from the illusion of substantial selfhood. Engler (1984) concludes: "You have to be somebody before you can be nobody" (p. 49). Engler's work makes a highly important contribution to the field of East–West studies by including a greater range of development than either psychotherapeutic or spiritual perspectives alone offers. Both psychoanalysis and Buddhism lack a full-spectrum psychology. The former has little to say about adaptive, non-self-centered states of subjectivity (which I shall explore in more detail in Chapters 3, 9, and 10) and psychological maturity and health. The latter neglects "earlier stages of personality organization and the types of suffering that result from a failure to negotiate them" (Engler, 1984, p. 49).

There is a tension in the developmental stage model between a complementary view of human development (one is first somebody and then nobody) and a complex, noncomplementary conception. In terms of the former, "Meditation and psychotherapy cannot be positioned on a continuum in any mutually exclusive way as though both simply pointed to a different range of human development. Not only do post-enlightenment stages of meditation apparently affect the manifestation and management of neurotic conditions, but this type of conflict continues to be experienced after enlightenment" (Brown & Engler, 1986, p. 212). In terms of the latter view, Brown and Engler conclude that "psychological maturity and the path to enlightenment are perhaps two complementary but not entirely unrelated lines of growth; or that they do represent different 'levels' or ranges of health/growth along a continuum, but with much more complex relationships between them than have previously been imagined" (p. 212).

Although Engler acknowledges that there are very complex interactions between conventional and contemplative stages and a "rigidly linear and unidirectional model is not at all what we have in mind" (Wilber, Engler, & Brown, 1986, p. 7), the complexity of interaction between "psychological" and "spiritual" perspectives is not addressed or spelled out. The limitations of contemplative perspectives and the value of conventional viewpoints are also neglected, especially the way the latter might enrich the former. Chapters 9 and 10 address this at greater length.

Transpersonal psychology was developed in the late 1960s by thinkers who felt that existing psychologies neglected the full range of human possibilities including transcendent states. Transpersonal psychology focuses on such things as altered states of consciousness and well-being, meditation, optimal psychological health, and the integration of therapeutic and spiritual disciplines. Wilber, Roger Walsh, Frances Vaughan, Stan Grof, and Charles Tart are some of its esteemed practitioners.

Wilber's spectrum psychology attempts to create a marriage between Western psychological perspectives on human development and psychopathology and Eastern contemplative understandings of consciousness and optimal states of health. His work exhibits encyclopedic scholarship, an exemplary groundedness in contemplative practices as well as theory, and an openness to diverse psychotherapeutic and spiritual traditions. In Wilber's work the quest to integrate Eastern contemplative and Western psychotherapeutic thought receives its most comprehensive and sophisticated expression.

Central to the spectrum of psychology is what Aldous Huxley (1944) has termed the *philosophia perennis*, the "perennial philosophy," a doctrine about the nature of humankind and reality underlying every major metaphysical tradition. It represents a "reality untouched by time or place, true everywhere and everywhen" (Wilber, 1979, p. 7). According to Wilber, corresponding to the perennial philosophy there exists a *psychologia perennis*, a "perennial psychology," a "universal view as to the nature of human consciousness, which expresses the very same insights as the perennial philosophy but in more decidedly psychological language" (1979, p. 7).

For Wilber, the crucial insight of the perennial psychology is that our "innermost consciousness is identical to the absolute and ultimate reality of the universe," which he terms "mind," which "is what there is and all there is, spaceless and therefore infinite, timeless and therefore eternal, outside of which nothing exists. On this level ... (one) is identified with the universe, the All—or rather ... is the All" (1979, p. 9). According to the perennial psychology this is "the only real state of consciousness, all others being essentially illusions" (1979, p. 9).

The perennial psychology is the foundation of Wilber's spectrum of consciousness model. The central underlying assumption of Wilbur's model is that "human personality is a multileveled manifestation or expression of a single consciousness, just as in physics the electromagnetic spectrum is viewed as a multibanded expression of a single, characteristic electromagnetic wave" (1979, p. 8). Consciousness, like light, exists on and is composed of various bands or spectrums, which develop through a series of stages that can be correlated with corresponding states of self-organization and self-blindness. Different psychological and spiritual traditions address these different levels.

In a recent article, Wilber (1986) proposes ten levels to the spectrum. In ascending order, they are: sensoriphysical, phantasmic—emotional, representational mind, rule—role mind, formal—reflexive mind, vision—logic, psychic, subtle, causal, and ultimate. It would distract from the central argument to define Wilber's terms. For our purposes it is sufficient to note that each stage of development has its own particular type of self-experience, cognitive development, moral sensibilities, potential distortions, and pathologies. Each level is characterized by a different sense of personal identity ranging from the narrow and circumscribed sense of identity, associated with the sensoriphysical level in which one identifies only with the realms of matter, sensation, and perception, to the ultimate level, in which one is identified with the totality of the universe. According to Wilber the great religious sages such as Buddha and the esteemed twentieth-century Hindu saint Ramana Maharshi are exemplars of the highest level of the spectrum.

On this level one is identified with the universe, the "All," or rather, is the All. This level is not an altered or abnormal state of consciousness but is "the only real state of consciousness, all others being essentially illusions" (Wilber, 1979, p. 9). One's "innermost consciousness is identical to the absolute and ultimate reality of the universe" (1979, pp. 8–9).

Each higher stage is less "self-centric" than its predecessors (Wilber, 1986). Each level can be correlated with corresponding ways of perceiving and misperceiving reality. Wilber maintains that different psychotherapeutic and spiritual traditions address and are best suited for different levels of the spectrum. Western psychotherapies (e.g., psychoanalysis, Gestalt therapy, and transactional analysis) address pathology and lower levels of the spectrum, while contemplative disciplines such as Buddhism are recommended for higher stages of the spectrum and the deepest kinds of transformation and liberation. For Wilber, psychoanalysis and Buddhism are complementary.

The value of Wilber's work, like Engler's (cf. 1984), is at least twofold: it disentangles meditative states of heightened clarity, health, and freedom from psychotherapeutic reductionism. Wilber and Engler maintain that contemplative practices constitute a higher and more advanced level of personality development "beyond ego" or the separate, autonomous, self-centered self that is the acme of mental health in most psychotherapies. Their second contribution is to offer guidance for meditators with psychological disturbances who are failing to make important discriminations in their meditation practice. Meditative practices, according to Wilber and Engler, may attract individuals with self-disorders, by which I mean people who experience themselves as brittle, fragile, worthless, vulnerable, and prone to self-esteem fluctuations. Meditators who experience self-issues of this sort (obviously not all meditators) may confuse their experiences of identity diffusion and depersonalization with genuine spiritual realization. For such individuals, Engler and Wilber recommend traditional therapy to shore up the self prior to pursuing meditation practice.

Psychoanalysis and Buddhism offer fertile possibilities for cross-pollination. Mutual enrichment, however, has been impeded by the restrictive perspective of previous studies that have adopted one of three monolithic viewpoints in characterizing their multifaceted relationship. These are what I would term the shotgun wedding, bridesmaid, and pseudo-complementary/token egalitarian models. I will briefly discuss each view before presenting my own alternative perspective.

Until relatively recently, much of the literature on Eastern and Western psychology has assumed either explicitly or implicitly that Buddhism and psychoanalysis are antithetical and incompatible. It is claimed that they occupy positions of unavoidable disagreement from which there can be no escape except by embracing one and abandoning the other (Rinzler & Gordon, 1980). Since psychoanalysis and Buddhism have very different visions of the mind and human existence, any attempt to join them is a "shotgun wedding" that "does justice to neither." A synthesis is thus "almost impossible" (Rinzler & Gordon, 1980, p. 52).

The most prevalent view of psychoanalysis and Buddhism is what I would term the bridesmaid stance in which psychoanalysis plays second fiddle to Buddhism. In the earlier literature, Buddhism was often subordinate to psychoanalysis (e.g., Alexander, 1931). When the bridesmaid perspective is operative, commerce between Buddhism and psychoanalysis occurs but only in one direction; in its more recent Orientocentric guise, writers emphasize Buddhism's value for psychotherapy (Boss, 1965; Chogyam, 1983; Deatherage, 1975) while neglecting the latter's value for Buddhism.

The third way that psychoanalysis and Buddhism have been approached, arguably the most compelling perspective, is Wilber's spectrum of consciousness model and Engler's developmental model. The spectrum model has tremendous theoretical and emotional appeal, since it promises to integrate apparently irreconcilable psychological and spiritual systems. Chaos seems to be reduced and seekers after truth no longer feel like United Nations delegates without an interpreter. A clinical vignette reveals certain problems with the spectrum model.

A 17-year-old boy becomes assailed by a fear of death. He loses all interest in worldly things. He is unable to concentrate. He does not do his schoolwork. For the past 2 or 3 months he has also become withdrawn, indifferent to family and friends. Family members become worried and angry. Teachers assign him extra homework in order to discipline him. One Sunday afternoon he is trying to do one such lesson but his attention wanders. His older brother walks in, sees his inattentiveness, and scolds him. The next day he runs away from home.

Previous habits of eating and sleeping go by the wayside. He sleeps rarely and erratically. He eats only if fed by concerned strangers. He stops speaking. In fact, he talks to no one for the next 4 years. He becomes oblivious, disheveled. He never bathes. Insects bite him, leaving pus-filled sores on his back and legs. He hardly notices.

This process in its acute form continues for close to a year. What has happened here? Unfortunately there is little in terms of a detailed history beyond this outline. Perhaps it would help to know that the boy's father died when he was 12 years old and that he had religious longings beginning rather suddenly at age 16. Specifically, a year or so before his departure from home, he had read a devotional book that stirred him to his depths. Thereafter, he visited the local temple every day for hours at a time, tears in his eyes, fervently praying to be made a true devotee of God. Upon fleeing home, he left a note that read, "I have started from this place in search of my Father in accordance with His command."

When an uncle and then his mother and brother first found him years later, he all but entirely ignored them. In his subsequent long life, he never worked at a job, never married, never developed normal relationships.

This young man, born Venkatarem Iyer, never did receive psychiatric treatment. Instead, he grew a bit older, settled down considerably, and came to be known as Sri Ramana Maharshi, one of the most deeply and universally admired saints in the history of India. For most anyone looking to the East for inspiration, Maharshi represents a phenomenal pinnacle of spirituality and wisdom. In fact, Maharshi is one of the sages Wilber cites as exemplifying the highest stage of mental health.

The spectrum model has several fundamental flaws. Development, according to this model, involves progressing through discrete and stratified stages ranging from disavowing aspects of one's identity to recognizing one's fundamental interconnectedness with everything. This presupposes, without actually demonstrating, that there is a uniformity to one's identity and stage of development and a separation and division of the psychological and spiritual.

The pathology of certain visionaries (Gordon, 1987; Schneider, 1987) and the prescience of some schizophrenics (Searles, 1972) teaches us that human functioning is much more complex than such schematic accounts suggest. One can experience the highest stage on Wilber's scale—unity consciousness—perceiving the interconnectedness of human existence, while also operating at times on "lower" levels, demonstrating myopia about one's body, feelings, or relationships. Some of the spiritual teachers embroiled in enormously egocentric and myopic behavior toward others around power, money, and sex demonstrate less interpersonal sensitivity and morality than people who are apparently operating on "lower" levels. One could also be operating on "lower" levels of the spectrum in certain areas while experiencing "higher" facets in other areas. I have worked with schizophrenics, for example, who struggle with the deepest kinds of self-disorders and have also at times perceived insights associated with "higher" levels of development on Wilber's model. They also have not treated others so capriciously and insensitively as the spiritual teachers who have manipulated others for their own benefit.

Because of the asymmetrical nature of human development, we all operate on different levels depending on which particular area of human experience and/or issues we are confronting. One could be quite aware of one's mental life and be disconnected from one's body, as some spiritual teachers and analysts are, or one could be attuned to one's body–mind and be relatively unaware of one's interpersonal relations and impact on others. The complexity and multidimensionality of human experience and development is obscured by linear, hierarchical, developmental models.

Wilber's model does not achieve genuine integration. The attempted "marriage" of psychological and spiritual perspectives is an asymmetrical affair in which Buddhism and other contemplative disciplines are actually viewed as superior to psychological thought, offering a privileged and true description of how humans really are. A tacit inequality is hidden underneath the nominal complementarity. There is an illusory rapprochement in which psychoanalysis and Buddhism are discreetly segregated to separate and unequal realms of reality and one is granted a special status. Whereas psychoanalysis usually pathologizes non-Western thought, transpersonal theorists sometimes romanticize it.

Within the transpersonal ranks, what I have recently termed "Orientocentrism" (Rubin, 1991, 1993) not Eurocentrism tends to predominate. Orientocentrism refers not to the "Orientalism" that literary and culture critic Edward Said (1979) critiques when he describes the tendency among Western commentators on the Orient to utilize an imperialistic discourse about Asia that fashions a distorted and reductionistic picture of "the East" in order to intellectually colonialize Asia and psychologically fortify itself. Rather, it refers to the mirror opposite danger to Eurocentrism: the valorizing and privileging of Asian thought and the neglect if not dismissal of Western psychological perspectives. The Zen master who told the student of Zen who indicated that psychotherapy and Zen had similar effects in overcoming suffering that the psychotherapist is just another patient (Matthiessen, 1987, p. 160) illustrates Orientocentrism, as does the absence of exploration concerning what value Western psychotherapies might have for non-Western thought in the preeminent, extant anthologies in the field of East–West studies [e.g., Welwood's (1979) *Meeting of the Ways*; Tart's (1975) *Transpersonal Psychologies*; Boorstein's (1980) *Transpersonal Psychotherapy*; Walsh & Vaughan's (1980) *Beyond Ego*]. Orientocentrism is so unconscious that no one has even remarked on its presence!

None of these perspectives on the relationship between the Western psychotherapeutic and Eastern contemplative disciplines—the shotgun wedding, bridesmaid approach or pseudo-complementary view—is wrong, but they reduce to a single factor or characterization what is a complex relationship with a multitude of dimensions. There are ways in which psychoanalysis and Buddhism are antithetical, complementary, *and* synergistic. But they are not simply any one of these all the time.

BEYOND EUROCENTRISM AND ORIENTOCENTRISM

I would like you to imagine for a few moments three types of couples. The first have nothing in common. They view the world and people differently; they seem to disagree about everything. The second couple have greater contact than the first one, but the relationship seems one-sided. One person, P is subordinated to the other, B. The one who dominates, B, seems to be the only partner who is capable of offering stimulation. Imagine a third couple. The one in power, B, claims that both members play complementary roles, each having an area of life in which they shine. But from the outside it is possible to observe that this is also a relationship of only nominal equality in which one member, usually P, is subordinated to B. None of these three types of relationship afford any genuine opportunity for intimacy.

In recent years there is much talk of a "dialogue" between psychoanalysis and Buddhism. In the spring of 1994, there was a conference at the Harvard Club in New York City entitled Healing the Suffering Self: A Dialogue among Psychoanalysts and Buddhists. Over 500 therapists and spiritual teachers and practitioners attended. As a participant, perhaps the most striking thing was the absence of contact between representatives of each tradition. It was evident that most of the time there was a monologue rather than a dialogue. Psychoanalysts and Buddhists spoke past each other rather than to each other. They did not seem to really question their own worldviews and to be open to what they might learn from each other. Rarely did anyone seem transformed by the encounter with the Other. Substituting psychoanalysis for "P" and Buddhism for "B" in the previous paragraph points to the three types of relatedness that have dominated the way people think about the relationship between Western therapeutic and Eastern contemplative disciplines. Each is an escape from intimacy that is a barrier to a relationship in which both are transformed and enriched by the encounter with the other.

"Truth," suggested Anatole France, "lies in the nuances." The nuances are exactly what the standard approaches to Western psychotherapies and Eastern contemplative disciplines neglect and eclipse. The relationship between Buddhism and psychoanalysis is more complex than the existing accounts suggest, forming not a singular pattern of influence but rather resembling a heterogeneous mosaic composed of elements that are—depending on the specific topic—antithetical, complementary *and* synergistic. For example, the goals of psychoanalysis and Buddhism are antithetical; the former focuses on strengthening one's sense of self, while the latter views such an enterprise as the very cause of psychological suffering. Meditative techniques for training attentiveness complement and enrich the psychoanalytic perspective on listening, while the psychoanalytic account of defense and resistance enhances the Buddhist understanding of interferences to meditation practice. Psychoanalytic and Buddhist strategies for facilitating transformation, as I shall demonstrate in the last two chapters, are, in some ways, synergistic.

The Eurocentrism of traditional Western psychology and the Orientocentrism of more recent writings on psychotherapeutic and contemplative disciplines both inhibit the creation of a contemplative therapeutics or an analytic meditation because they establish an intellectual embargo on commerce between Asian and Western psychology. An alternative perspective is necessary for the genuine insights of each tradition to emerge. In contrast to the Eurocentrism of psychoanalysis and the Orientocentrism of much recent discourse on psychoanalysis and Buddhism, I will be recommending a more egalitarian relationship in which there is mutual respect, the absence of denigration or deification and submission or superiority, and a genuine interest in what they can teach each other.

The egalitarian relationship I am pointing toward is not meant to be a complementarity that erases differences or subsumes either psychoanalysis into Buddhism or Buddhism into psychoanalysis in the act of detecting similarities. Since the advent of deconstructionism, the limitations of searching only for commonalities between two systems of thought appears more problematic. It misses what is most interesting, which is how they are different, what the common denominators eclipse, and how both systems are incompatible.

If we abandon the three traditional approaches to psychoanalysis and Buddhism, we shall cease asking questions like "Which is true?" or "Which is better?" and instead ask additional questions such as "What do Buddhism and psychoanalysis evade?" "What can psychoanalysis teach Buddhism?" and "What can Buddhism teach psychoanalysis?" In the next chapter, I will explore these questions in relation to psychoanalytic and Buddhist views of self.

REFERENCES

Alexander, F. (1931). Buddhist training as an artificial catatonia. *Psychoanalytic Review*, *18*, 129–145.

Assagioli, R. (1971). *Psychosynthesis*. New York: Viking.

Bion, W. (1970). *Attention and interpretation*. New York: Basic Books.

Boorstein, S. (Ed.). (1980). *Transpersonal psychotherapy*. Palo Alto, CA: Science and Behavior Books.

Boss, M. (1965). *A psychiatrist discovers India*. London: Oswald Wolff.

Brown, D., & Engler, J. (1986). The stages of mindfulness meditation: A validation study. Part II: Discussion. In K. Wilber, J. Engler, & D. Brown (Eds.), *Transformation of consciousness: Conventional and contemplative perspectives on development* (pp. 193–216). Boston: Shambhala.

Chogyam, T. (1983). Introductory essay. In N. Katz (Ed.), *Buddhist and Western psychology* (pp. 1–7). Boulder, CO: Prajna Press.

Cohen, P. (1986). *An exploratory study of religiously committed psychoanalytically oriented clinicians*. Unpublished doctoral dissertation, City University of New York.

Deatherage, G. (1975). The clinical use of "mindfulness" meditation techniques in short-term psychotherapy. *Journal of Transpersonal Psychology*, *7*, 133–143.

Engler, J. (1984). Therapeutic aims in psychotherapy and meditation: Developmental stages in the representation of self. In K. Wilber, J. Engler, & D. Brown (Eds.), *Transformations of consciousness: Conventional and contemplative perspectives on development* (pp. 17–51). Boston: Shambhala, 1986.

Finn, M. (1992). Transitional space and Tibetan Buddhism: The object relations of meditation. In M. Finn & J. Gartner (Eds.), *Object relations theory and religion* (pp. 87–107). Westport, CT: Praeger Press.

Freud, S. (1900). The interpretation of dreams. In J. Strachey (Ed. & Trans.), *The standard edition of the complete psychological works of Sigmund Freud*, Vols. 4 & 5. London: Hogarth Press.

Freud, S. (1926). Inhibitions, symptoms and anxiety. In J. Strachey (Ed. & Trans.), *The standard edition of the complete psychological works of Sigmund Freud*, Vol. 20 (pp. 77–175). London: Hogarth Press.

Freud, S. (1927). The future of an illusion. In J. Strachey (Ed. & Trans.), *The standard edition of the complete psychological works of Sigmund Freud*, Vol. 21. London: Hogarth Press.

Freud, S. (1930). Civilization and its discontents. In J. Strachey (Ed. & Trans.), *The standard edition of the complete psychological works of Sigmund Freud*, Vol. 21. London: Hogarth Press.

Freud, S. (1933). New introductory lectures. In J. Strachey (Ed. & Trans.), *The standard edition of the complete psychological works of Sigmund Freud*, Vol. 22 (pp. 5–182). London: Hogarth Press.

Fromm, E., Suzuki, D.T., & DeMartino, R. (Eds.). (1970). *Zen Buddhism and psychoanalysis*. New York: Harper & Row.

Gay, P. (1987). *A Godless Jew*. New Haven, CT: Yale University Press.

Goleman, D. (1988). *The meditative mind*. New York: Tarcher.

Gordon, J. (1987). *The golden guru: The strange journey of Bhagwan Shree Rajneesh*. Lexington, KY: The Stephen Greene Press.

Group for Advancement of Psychiatry. (1968). *The psychic function of religion in mental illness and health* (Report 67). New York: GAP.

Horney, K. (1945). *Our inner conflicts*. New York: Norton.

Horney, K. (1987). *Final lectures*. New York: Norton.

Huxley, A. (1944). *The perennial philosophy*. New York: Harper and Row.

Jones, E. (1957). *The life and work of Sigmund Freud*, Vol. 3. London: Hogarth Press.

Jung, C.G. (1929). *Collected works, Vol. 13*. Princeton, NJ: Princeton University Press.

Jung, C.G. (1952). *Collected works, Vol. 5*. Princeton, NJ: Princeton University Press.

Jung, C.G. (1958). *Collected works, Vol. 11*. Princeton, NJ: Princeton University Press.

Jung, C.G. (1968). *Collected works, Vol. 12*. Princeton, NJ: Princeton University Press.

Kelman, H. (1960). Psychoanalytic thought and Eastern wisdom. In J. Ehrenwald (Ed.), *The history of psychotherapy*. New York: Jason Aronson.

Kohut, H. (1985). *Self psychology and the humanities*. New York: Norton.

Lovinger, R. (1989). *Religion in the stabilization and regulation of the self*. Unpublished manuscript.

Maslow, A. (1971). *The farther reaches of human nature*. New York: Viking Press.

Masson, J.M., & Masson, T.C. (1978). Buried memories on the acropolis; Freud's response to mysticism and anti-Semitism. *International Journal of Psychoanalysis, 59*, 199–208.

Matthiessen, P. (1987). *Nine-headed dragon: Zen journals 1969–1982*. Boston: Shambhala.

Meissner, W. (1984). *Psychoanalysis and religious experience*. New Haven, CT: Yale University Press.

Meng, H., & Freud, E. (Eds.). (1963). *Psychoanalysis and faith: The letters of Sigmund Freud and Oskar Pfister*. New York: Basic Books.

Menninger, K. (1942). *Love against hate*. New York: Harcourt, Brace, Jovanovich.

Milner, M. (1973). Some notes on psychoanalytic ideas about mysticism. In M. Milner, *The suppressed madness of sane men* (pp. 258–274). London: Tavistock.

Nyanaponika, T. (1962). *The heart of Buddhist meditation*. New York: Samuel Weiser.

Nyanaponika, T. (1972). *The power of mindfulness*. San Francisco: Unity Press.

Pfister, O. (1948). *Christianity and fear* (W.H. Johnston, Trans.). London: George.

Pruyser, P. (1971). Assessment of the psychiatric patient's religious attitudes in the psychiatric case study. *Bulletin of the Menninger Clinic, 35,* 272–291.

Pruyser, P. (1973). Sigmund Freud and his legacy: Psychoanalytic psychology of religion. In C.Y. Block & P.E. Hammond (Eds.), *Beyond the classics* (pp. 243–290). New York: Harper & Row.

Quinn, S. (1987). *A mind of her own: The life of Karen Horney.* New York: Summit Books.

Rinzler, C., & Gordon, B. (1980). Buddhism and psychotherapy. In G. Epstein (Ed.), *Studies in non-deterministic psychology* (pp. 52–69). New York: Human Sciences Press.

Rizzuto, A. (1979). *The birth of the living God.* Chicago: University of Chicago Press.

Roland, A. (1988). *In search of self in India and Japan: Toward a cross-cultural psychology.* Princeton, NJ: Princeton University Press.

Rubin, J.B. (1985). Meditation and psychoanalytic listening. *Psychoanalytic Review, 72*(4), 599–613.

Rubin, J.B. (1991). The clinical integration of Buddhist meditation and psychoanalysis. *Journal of Integrative and Eclectic Psychotherapy, 10*(2), 173–181.

Rubin, J.B. (1992). Psychoanalytic treatment with a Buddhist meditator. In M. Finn & J. Gartner (Eds.), *Object relations theory and religion: Clinical applications* (pp. 87–107). Westport, CT: Praeger.

Rubin, J.B. (1993). Psychoanalysis and Buddhism: Toward an integration. In G. Stricker & J. Gold (Eds.), *Comprehensive textbook of psychotherapy integration* (pp. 249–266). New York: Plenum Press.

Saffady, W. (1976). New developments in the psychoanalytic study of religion: A bibliographic review of the literature since 1960. *Psychoanalytic Review, 63*(2), 291–299.

Said, E. (1979). *Orientalism.* New York: Vintage Books.

Schneider, K. (1987). The deified self: A "centaur" response to Wilber and the transpersonal movement. *Journal of Humanistic Psychology, 27*(2), 196–216.

Searles, H. (1972). The function of the patient's realistic perceptions of the analyst in delusional transference. In *Countertransference and related subjects* (pp. 196–227). New York: International Universities Press.

Silberer, H. (1917). *Problems of mysticism and its symbolism.* New York: Moffat.

Suler, J. (1993). *Contemporary psychoanalysis and Eastern thought.* Albany, NY: State University Press.

Suzuki, D.T. (1974). *An introduction to Zen Buddhism.* New York: Causeway.

Tart, C. (Ed.). (1975). *Transpersonal psychologies.* New York: Harper & Row.

Walsh, R., & Vaughan, F. (Eds.). (1980). *Beyond ego: Transpersonal dimensions in psychotherapy.* Los Angeles, CA: Tarcher.

Watts, A. (1961). *Psychotherapy East and West.* New York: Pantheon Books.

Welwood, J. (Ed.). (1979). *Meeting of the ways: Explorations in East/West psychology.* New York: Schocken.

Wilber, K. (1979). Psychologia perennis. In J. Welwood (Ed.), *Meeting of the ways: Explorations in East/West psychology.* (pp. 7–28) New York: Schocken.

Wilber, K. (1983). *A sociable God.* New York: McGraw Hill.

Wilber, K. (1986). The developmental spectrum and psychopathology; Part I, Stages and types of pathology. *Journal of Transpersonal Psychology, 16*(1), 75–118.

Wilber, K., Engler, J., & Brown, D. (1986). *Transformations of consciousness: Conventional and contemplative perspectives on development.* Boston: Shambhala.

Winnicott, D.W. (1986). *Home Is Where We Start From.* New York: Norton.

I

Psychoanalytic and Buddhist Conceptions of Self

3

Beyond Self-Blindness
Psychoanalytic and Buddhist Visions of the Self

The aspect of things that are most important to us are hidden because of their simplicity and familiarity. One is unable to notice something because it is always before one's eyes. The real foundation of ... [one's] inquiry do not strike ... a [person] at all.... And this means: we fail to be struck by what, once seen, is most striking and most powerful.
—Ludwig Wittgenstein

One must look for the assumptions which appear so obvious that people do not know they are assuming them because no other way of putting things has ever occurred to them.
—Alfred North Whitehead

The greatest danger, according to the nineteenth-century Danish philosopher Sören Kierkegaard, is to lose one's self. And the greatest despair is not being one's self. But what is this self that one can lose or not be?

The nature of the self is of central concern to both psychoanalysis and Buddhism: "To study the Way [Zen]" notes Zen master Dogen (1200–1253), "is to study the self." The self, according to psychoanalyst Heinz Kohut (1984), is the organizing center of the individual's psychological universe. Although differing significantly in theory and practice, both psychoanalysis and Buddhism explore subjectivity with lapidary precision and share an assumption that optimal development involves a transformation of the self. But, their differing viewpoints about how to investigate the self and what it is lead in diametrically opposed theoretical and clinical directions.

Consider the following two vignettes, the first Buddhistic and the second psychoanalytic.

In an unreal green forest you are walking next to the old master. You get to a brook. The master touches your shoulder and you know that he wants you to sit down.... He shakes his head and points at a piece of cork floating past; it

57

has been in a fire and half of it is black. "That piece of cork is your personality," the master says. "At every turn, at every change of circumstances, at
every conflict, defeat or victory, a piece of it crumbles off." You look at the
piece of cork. Pieces of it detach themselves and disappear. The cork is
getting smaller. "It is getting smaller," you say, nervously. "Getting smaller
all the time." The quiet voice of the master is very close.... "Till nothing is
left of it." ... He looks quiet and pleasant. There is only a little old man who
wants to point something out to you.... You will lose your name, your body,
and your character. Your fear diminishes. If it has to happen, it will happen.
Nothing will remain. And nothing you will be. (van de Wetering, 1975, pp.
21–22)

 James, a twenty-eight-year-old chemist, continually maintained that he
could not become a "person" and had "no self": "I am only a response to
other people, I have no identity of my own. I am only a cork floating on the
ocean." (Laing, 1965, pp. 47–48)

James did not experience the affirmative and expansive sense of
nothingness Jan van de Wetering depicts in a Japanese Zen monastery.
In fact, James' sense of nonbeing—his experience of ontological weight-
lessness and nonexistence—seemed to be more of a living hell than an
exalted nirvana. But the Zen student's experience should not be dismissed by psychoanalysts nor the analysand's by Buddhists without
further investigation because they each raise important questions for the
other that ordinarily would remain unasked if one pursued either psychoanalysis or Buddhism in isolation.

|If preoccupation with the self arises, as Harry Stack Sullivan (1950)
suggests, in order to lessen anxiety and promote self-security—"The
self-system ... is an organization of experience for avoiding increasing
degrees of anxiety" (p. 166)—is the self-preoccupied, individualized
self that pervades Western civilization an epiphenomenon of the mal-
adaptive and anxiety-generating ways of Western life?|Would the elim-
ination of some of the irrational and antihuman ways of being that are
encouraged by Western societies result in a different, less self-centric
self-system? Would self-centricity not arise in a person who did not
experience anxiety?

|If the self is, as Roy Schafer (1989) maintains, a "narrative construc-
tion" (p. 165)—the collection of narratives one creates about oneself—
which may include stories about the false self and the true self, the
imprisoned self and the questing self, the empty self, the reborn self, the
maligned self, the guilty self, the constrained and the contented self and
so on, is the individualized self a Westerncentric narrative?|Does the
Zen student's experience embody a state of self-transcending subjectivity? Does the student of Zen experience a healthy state of subjectivity
beyond what Western psychoanalysts have mapped? What does the
healthy self look like after a successful analysis? Is there a transcen-

dence of narcissistic subjectivity? Is that different from the experience of the Zen student?[1]

Contrariwise, psychoanalysis raises a series of important questions for Buddhism: Why are there so many documented instances in recent years of grossly self-centered and exploitative behavior among Buddhist teachers in the United States? (cf. Boucher, 1988). Given the fact that psychological conditioning and structure may be inevitable, are there any dangers and self-deceptions involved in the search for self-transcendence? Are there any dangers in Buddhism's encouraging of self-nullification? How does Buddhism deal with self-unconsciousness and stable and enduring patterns of thought, action, and behavior in its model of subjectivity? What does Buddhism omit from its cartography of subjectivity?

When I refer to Buddhist meditation, I shall be focusing on Vipassana meditation, its core practice. There are two types of meditation practices in Buddhism: concentration and Vipassana or insight. The former involves attending to the experience of one phenomenon, such as the sensation of breathing, until the mind becomes focused and concentrated. Vipassana is a practice of a "clear and single-minded awareness of what actually happens to us and in us at the successive moments of perceptions" (Nyanaponika, 1973, p. 30). One attends, without selection or judgment, to the experience of whatever mental or physical phenomena, for example, thoughts, feelings, sensations, or fantasies, are predominant in the field of awareness. This practice cultivates "mindfulness," or refined nonselective and nonjudgmental awareness of whatever is occurring, which eventually leads to highly refined perceptual acuity and attentiveness, increased control of apparently voluntary processes, deepened insight into the nature of mental and physical processes, selfhood and reality, the lessening of suffering, and the development of exemplary compassion and moral action.

The sense as you read this paragraph that you exist as a person; that there is something correct to call "me"; something that has its own body, feelings, fantasies, goals, dreams, pain, and ideals—the sense that I am a self—is, for most psychoanalysts (with the exception of Lacanians), a taken-for-granted reality, perhaps the linchpin of human life.

The reality of subjectivity is built into the very fabric of the English language and our daily lives. The English language, as Schafer (1989) notes, "authorizes us to think and speak in terms of single, stable self-entities.... Locutions such as "be yourself" and "divided self" are in-

[1]Discussions with Stephen Mitchell opened up some of these questions and sharpened my attempts to examine psychoanalytic and Buddhist views of subjectivity in tandem.

stances of what I mean" (p. 159). Psychoanalysts take for granted the self
as "the center of initiative and identity, who is the source of needs and
desires, the one who feels and thinks and acts, the defender and execu-
tor" (Meyers, 1989, p. 139).

With rare exceptions they also take for granted the universal value
of the concept of the self. Whether or not the disconnection and anomie
that permeates the experience of late twentieth-century persons is in
any way a result of a culturally sanctioned neurotic preoccupation with
the sovereignty and separateness of our selves is a question that it might
be instructive to investigate.

The fate of this self, according to Jack Engler (1986), a psychoanalytically informed psychologist who has engaged in the most extensive psychological study of advanced meditation practitioners from
America and Southeast Asia, which I will discuss in Chapter 4, is an
issue on which psychoanalytic and Buddhist perspectives

> seem diametrically opposed.... The deepest psychopathological problem
> from the perspective of [psychoanalysis] ... is the lack of a sense of self. The
> most severe clinical syndromes—infantile autism, the symbiotic and func-
> tional psychoses, the borderline conditions—are precisely failures, arrests
> or regressions in establishing a cohesive, integrated self or self concept. In
> varying degrees of severity, all represent disorders of the self, the inability to
> feel real or cohesive or "in being" at all. In contrast, the deepest psycho-
> pathological problem from the Buddhist perspective is the presence of a self
> and the feeling of selfhood. According to Buddhist diagnosis, the deepest
> source of suffering is the attempt to preserve a self, an attempt which is
> viewed as both futile and self-defeating. The severest form of psychopathol-
> ogy is precisely attavadupadana, the clinging to personal existence. (Engler,
> 1986, pp. 23–24)

The self, as it is ordinarily perceived, is an illusion, according to
Buddhism, an epiphenomenon generated by our perceptual insensitivity. We imagine there is a unified, independent self because without
special perceptual training, such as meditation, we view subjectivity
grossly. Like nonphysicists attempting to examine the particles of a table
without the benefit of an electron microscope, we do not see the fluidity,
discontinuity, and insubstantiality of subjectivity underneath the apparent solidity, continuity, and substantiality.

The consensually accepted belief, prevalent in most sectors of
Western life, in the substantiality of the illusion of substantial selfhood
is, according to Buddhism, problematic in several ways: (1) it presents a
distorted view of how things really are; (2) it interferes with experiencing a deeper, more profound reality; and (3) it generates suffering and
interferes with experiencing the greatest kind of contentment (cf. Nozick,
1989). Attachment to personal existence is eschewed by Buddhists be-

cause it is impermanent, it lacks enduring substance, and it cannot satisfy even the simplest of desires (cf. Engler, 1986).

Understanding the absence of an abiding self leads, according to Buddhism, to an absence of self-centric preoccupations and a more engaged relationship to life. This leads to greater compassion and a sense of contentment and openness. Rather than focusing on the development of the self, Buddhism recommends and attempts to facilitate seeing through its illusoriness.

The Buddhist perspective on selfhood is illustrated by a conversation an Indian teacher of Buddhist meditation had with one of her students, an American teacher of Buddhism, when she was visiting the United States. After leaving an automated bank machine she looked slightly sad and disturbed. Her student asked her what was wrong. She said that she felt badly for the person who had to work all day in cramped quarters behind the machine processing the transactions. Her student informed her that there was no one behind the wall; that the transactions occurred automatically. The Indian teacher said that was like the Buddhist conception of the self: processes occur—wishing, thinking, seeing, and so forth—but there is no single, autonomous, self-identical person performing them (Joseph Goldstein, personal communication).

A cohesive and integrated self, which is generally recognized as the acme of human development by most psychoanalysts, is viewed by Buddhists as a "state of arrested development" (Engler, 1986, p. 47) that is a basic and inevitable source of psychological pain and suffering. In pointing to psychological life "beyond ego," Buddhism is positing a level of self-development and health that is omitted from and thus foreign to the psychoanalytic diagnostic spectrum.

Believing that the self is an illusion, a Buddhist would view James' quest in therapy to develop an identity as a misguided enterprise that would be a source of enormous psychological suffering. Buddhism would attempt to facilitate James' radical transcendence of self-concern. In contrast, a psychoanalyst, depending on his or her theoretical orientation, would probably attempt to help James develop a sense of integration and identity, including a sense of self-cohesion, temporal stability and agency, and a set of personally chosen values and ideals. A psychoanalyst would probably view Buddhism's denial of selfhood as some sort of symptom of conflict or developmental arrest that needed to be understood and worked through psychoanalytically. Psychoanalytic and Buddhist views of the nature of subjectivity do indeed seem, as Engler (1986) claims, "diametrically opposed."

"To get clear about philosophical problems," suggests Ludwig Wittgenstein (1958), "it is useful to become conscious of the apparently unimportant details of the particular situation in which we are inclined to make a certain metaphysical assertion" (p. 66). Perhaps the most important detail of the particular situation in which psychoanalysts and Buddhists theorize about subjectivity is the context of investigation in which psychoanalysts and Buddhists examine subjectivity, which previous comparative studies of psychoanalysis and Buddhism (e.g., Engler, 1986; Epstein, 1989, 1990, 1995; Schuman, 1991) never examine. They resemble, in what might be termed their pre-Heisenbergian neglect of the context that informs their investigations of subjectivity, an archeologist who "discovers" a wristwatch that he or she unknowingly drops into a dig (cf. Stolorow & Lachmann, 1984/5).[2]

Psychoanalysts and Buddhists examine subjectivity from radically different vantage points that lead to very different conceptions of it. I shall highlight three aspects of their differing methodologies.

First, subjectivity can be investigated from a telescopic/wide-angle or a microscopic/zoom lens approach (Nozick, 1989). The latter mode of investigation refers to the attempt to capture a given moment of experience and examine it microscopically. When we examine our experience in this way we tend to discover, as William James (1890) noted, nothing inner or active—only a variety of fleeting thoughts, feelings, perceptions, or kinesthetic sensations. The former approach involves a more generalized, wide-angle, and unfocused mode of introspection that might resemble gazing at the ocean without focusing on any particular aspect of it such as a wave.

Consciousness, in James' (1890) view, like the journey of birds, takes two forms: movement, or the flight, and stasis, or the perching. Introspection, especially in its wide-angle form, is particularly prone to the prejudice of overemphasizing what James terms the "substantive" aspects of experience, the perchings of the bird, while neglecting what James terms the "transitive," the flight of the bird/the feelings of relationship or activity. James compares trying to see the transitive dimension with attempting to catch a snowflake crystal. The act of trying to grasp it annihilates it.

---

[2]From Sullivan's (1940) "participant–observer" to Langs' (1976) "bipersonal field" to Atwood and Stolorow's (1984) "intersubjectivity" theory to Mitchell's (1983) "relational matrix," many analysts have recognized both that the psychoanalytic field is shaped by the input of the analyst and that the psychoanalytic investigator influences the nature and the results of the investigation. Unfortunately, these insights have been applied to patient's communications (Langs, 1976), transference (Stolorow & Lachmann, 1984/5), and the psychoanalytic environment (Atwood, & Stolorow, 1984), rather than psychoanalytic conceptions of subjectivity.

The distinction biologists make between the "proximate" or the short-term causes of phenomena and the "ultimate" or long-term causes of phenomena (Modell, 1985, p. 85) provides the second difference between psychoanalytic and Buddhist examinations of subjectivity. Buddhists focus on the former, while psychoanalysts explore both the former and the latter.

Philosopher Richard Wollheim's (1984) distinction between episodic "mental states," such as thoughts, moments of interest, boredom, joy, lust, despair, and so forth, and recurrent "mental dispositions," which do not occur and are not events, such as "knowledge and belief, emotions, desires, habits, virtues and vices, and skills" (pp. 33–34) provides the third dimension to the different contexts in which Buddhists and psychoanalysts examine subjectivity. Buddhist meditation is designed to examine mental states, while psychoanalysis tends to place greater emphasis on mental dispositions. In the next section I will elaborate on these issues.

PSYCHOANALYTIC VIEWS OF SUBJECTIVITY

The analyst's office is the laboratory in which psychoanalytic data about subjectivity emerge. The analyst listens with "evenly hovering attention" to the patient's "free associations." What the analyst listens to is apprehended within a particular hermeneutical context, a genetic, linear perspective (Gold, 1992) in which it is assumed that the particular symptoms, mental states, traits, or conduct in the present have been shaped in a linear way by formative experiences in the past.

Psychoanalysts pay attention to the proximate or instantaneous manifestations of the analysand's material including episodic "mental states." However, what Freud termed the attraction of the infantile prototypes, which includes recurrent "mental dispositions," may lead to an overemphasis on the ultimate aspects of subjectivity and a slight neglect of some of the "here-and-now" aspects of the person (Gill, 1982).[3] This will be discussed in more detail later in the chapter.

Through this kind of wide-angle attention to the patient's material the analyst "discovers" a substantial, enduring subject shaped by formative experiences and events from the long-term past. The typology of this substantial self varies depending on the school of psychoanalysis. In his synthetic overview of postclassical psychoanalytic views of the self, Mitchell (1991) suggests that each school of psychoanalysis con-

---

[3]This is not true of more interpersonally oriented analysts.

strues selfhood differently. Mitchell distinguishes three views of the self: the Freudian view of the self as separate and integrated, the object relations and interpersonal view of it as multiple and discontinuous, and the self psychological view of it as integral and continuous. Mitchell (1991) maintains that the

> portrayal of self as multiple and discontinuous and the sense of self as separate, integral and continuous are referring to different aspects of self. The former refers to the multiple configurations of self patterned variably in different relational contexts. The latter refers to the subjective experience over time and across the different organizational schemes. The experience of patterning may be represented as having particular qualities or tones or content at different times; however, at every point, it is recognized as "mine," my particular way of processing and shaping experience. (p. 139)

But despite these differences there is a general agreement among psychoanalysts about two things: (1) the self exists, and (2) strengthening and expanding it is a fundamental goal of psychoanalysis. The outcome of psychoanalysis is an expanded and nuanced experience and understanding of "I-ness."

In a reflection late in his life on the nature of self-experience, Harry Stack Sullivan (1950) wondered if it were possible for there to be a completely different sort of self-system: "for all I know, there would not be evolved, in the course of becoming a person, anything like the sort of self-system that we always encounter" (p. 168). Sullivan ultimately rejected his intriguing question, maintaining that a human being without a self-system is "beyond imagination" (p. 168). But it is not beyond a Buddhist imagination.

BUDDHIST APPROACHES TO SUBJECTIVITY

The laboratory in which Buddhist views of subjectivity are formulated is the practice of meditation, which can occur either in a retreat context removed from the busyness and sensory stimulation of daily life or in the midst of everyday life. Meditation is the careful and detailed observation of the moment-to-moment, proximate dimensions of consciousness, for example, mental states such as the feeling or thought that just arose as you read this. One attempts to attend to whatever occurs with nonselective and nonjudgmental awareness. One strives to treat lapses in attentiveness and judgments about such imperfections in the same manner.

Experimental research findings demonstrate that this process cultivates enhanced perceptual and introspective sensitivity and acuity (cf.

Walsh, 1989). Preliminary findings of tachistoscopic research under-taken by Dr. Dan Brown of Cambridge Hospital, Harvard Medical School, on advanced Vipassana practitioners, which studies high-speed information processing prior to conscious attention, confirms "a percep-tual discrimination capacity well beyond hitherto reported norms, and tend to support the hypothesis that meditators are actually discriminat-ing temporal stages of high-speed processing prior to the build-up of stimuli into durable percepts" (quoted in Engler, 1986, p. 317, footnote 13). Through this kind of microscopic, "zoom lens" (Nozick, 1989) attention to consciousness, Buddhism "discovers" an ever-changing flux of thoughts, feelings, perceptions, memories, and sensations that arise and pass away like clouds entering and exiting from the sky.

There is no single, neutral, objective, ahistorical, transcendental viewpoint from which to view the self. Our conception of subjectivity depends on *how* we look at it, which usually remains unconscious, and thus unexamined by psychoanalysts and Buddhists.

What psychoanalysts and Buddhists "discover" about selfhood is often an artifact of the way they investigate it. The solidity and continu-ity of a unified subject shaped by a particular history of recurrent self and object images and associated affects and mental dispositions emerges for a psychoanalyst approaching subjectivity with a wide-angle lens, while the proximate nature of subjectivity, the way we are shaped and partially created anew each moment by the arising of fluid and hetero-genous mental states, is evident to a Buddhist utilizing a microscopic perspective.

As a result of their different observation points in investigating subjectivity, psychoanalysis and Buddhism each neglect different facets of it. Each conception fosters a complementary type of self-blindness. In the next section I shall discuss how Buddhists are usually "nearsighted" about subjectivity and psychoanalysts are often "farsighted." Buddhists neglect the substantial, enduring historical aspects of it. Questions of the subject's history and agency are eclipsed in Buddhism and the seeds of self-blindness are thus planted. In examining subjectivity psycho-analytically, psychoanalysis is sometimes "farsighted" about it, neglect-ing several of its near-at-hand aspects.

MUCH ADO ABOUT NOTHINGNESS OR EVADING THE SUBJECT

Buddhism has an atomistic "perceptual level bias" (Nicholson, 1989, p. 1). It emphasizes paying attention to the most "infinitesimal building blocks of perceptual events" (p. 1). Like subatomic physics, it focuses on the microscopic aspects of experience. The tacit assumption seems to be that the further one can reduce something into its constituent parts, then the more real they are and the better one understands them. Paying attention to microscopic phenomena, according to this perspective, is superior to a more macroscopic awareness.

What is rarely recognized by Buddhists is that investigating subjectivity in this way profoundly and irrevocably shapes and delimits what is noticed. More specifically, it creates an erroneous, self-nullifying image of human subjectivity.

Buddhists are unaware of what William James recognized about the inherent elusiveness of investigating transitive aspects of human consciousness. As I suggested previously, James compared trying to see the transitive dimension of human subjectivity with the inherently self-defeating attempt to catch a snowflake crystal.

Those who suggest that the ineffability of transitive phenomena proves their nonexistence resemble, in James' (1890) view, "Zeno's treatment of the advocates of motion, when asking them to point out in what place an arrow is when it moves, he argues the falsity of their thesis from their inability to make to so preposterous a question an immediate reply" (p. 244). The Buddhist search for the essence of subjectivity with a microscopic methodology thus may exemplify the process Wittgenstein (1958) described as "In order to find the real artichoke, we divested it of its leaves" (I, sec. 164).

The atomistic orientation of Buddhism is necessarily ahistorical. The absence of a sense of history or personal continuity leaves one adrift without any anchor. Jorge Luis Borges' (1970) frightening story in *Labyrinths* of the man for whom the world was so particular that he could not think or generalize typifies the dangers inherent in Buddhism's ahistorical, self-nullifying viewpoint. "Not only was it difficult for him to comprehend that the generic term dog embraces so many unlike individuals of diverse size and form, it bothered him that the dog at three fourteen (seen from the side) should have the same name as the dog at three fifteen (seen from the front)." The world is a living nightmare for this man, with no reference point to make sense of experiences (cf. Wood, 1989).

In throwing out the bathwater of egocentricity, Buddhism eliminates the baby of human agency and may preclude political engagement. If there is no subject, then there is no agent to evaluate phenomena and no previous experience from which to learn. There is thus no one who is exploited or alienated and no oppression to challenge or contest. Furthermore, there would appear to be no basis for morality or effective social criticism when human experience is viewed as a set of discon-

tinuous serial moments.[4] There are no established standards to adjudi-
cate between various courses of action and no criteria for decisions or
actions. Crucial questions about resistance to the existing order of society are eclipsed: Resistance in the name of what? For the sake of whom?
To what end? (Walzer, 1988, p. 191). Meditators will have greater difficulty, in certain ways, leading a full life in the present and creating an
individualized array of goals and ideals than they would have if they did
not deny their status as historical beings.

One consequence of this is that it helps people not to be critical of
things they should be critical of. This can result in a self-induced historical "amnesia" in which amoral actions in spiritual communities, for
example, can be rationalized as things in the past that are of no concern
because they are not happening in and do not pertain to the present
moment.

Evading the subject also leads to the "return of the repressed." Self-
denial leads to self-centeredness. The result of denying the existence of
a substantial self is a resurfacing of some of the disavowed aspects of
subjective life. In recent years there has been an epidemic of scandals in
American Buddhist communities involving Buddhist teachers illegally
expropriating money from the community or sexually exploiting students (Boucher, 1988). The acting out of such self-centered behavior is,
in my view, directly related to Buddhism's denial of self-existence. Such
incidents could only happen when there is a shocking and self-disabling
unselfconsciousness. The return of the repressed emerges in this acting
out on the part of Buddhist teachers. Those who deny the existence of at
least some self-centeredness are condemned to self-centeredly enact it.
Thus, instead of a meditation teacher utilizing the feeling of sexual
attraction to a student as feedback about important personal and perhaps interpersonal phenomena, these feelings may be denied and disavowed and then acted out. It is thus not incidental or accidental that
there have been so many incidents of self-centered behavior in spiritual
communities espousing a self-denying ethos in the United States in
recent years. In focusing on the personal aspects of this phenomena I do
not mean to omit the structural facets, the corruption shaped by what
Kramer and Alstad (1993) term the "authoritarian–hierarchical" relationship of teacher and student, which inevitably leads to self-sub-

---

[4]With its commitment to nonharming and its emphasis on diminishing self-centered and
exploitative cognition and conduct, Buddhism can be viewed as an ethical psychology.
It provides a highly elaborated ethical framework and praxis applicable to a range of
human activities including interpersonal relations, speech, work, and sexuality. My
point is that Buddhist theorizing on human subjectivity taken to its logical conclusion
has certain rarely articulated and ethically disabling implications.

mission, self-diminishment, self-blindness, and corruption. I shall discuss this, as well as other aspects of this phenomenon, in Chapter 9.

"FIXING" THE SUBJECT

But if Buddhism evades the subject, psychoanalysis tends to "fix it"—to cut off some of its life. The psychoanalytic approach to subjectivity leads to "farsightedness."

Psychoanalysis often fails to recognize the open and perpetually unfolding nature of identity. The impact on consciousness and conduct in the present of episodic, transient mental states that arise moment after moment and may not fit neatly into or confirm our established, genetically based narrative of ourselves is usually neglected in psychoanalysis. It also does not recognize or value the nonpathological, non-self-centric aspects of subjectivity.

That psychoanalysis has often been guilty of a fixed conception of identity is illustrated by its predilection for reifying human subjectivity. Self-reification emerges in the tension in various schools of psychoanalysis, whether classical, object relations theory, or self psychology, between techniques of investigating subjectivity that encourage the emergence of the uniqueness of patient's self-experience and theories of subjectivity that essentialize it such as Freud's tripartite self, Winnicott's true self, and Kohut's bipolar self (cf. Rubin, 1995). Self-reification denies both the open and continuously unfolding quality of human identity and the non-self-centric facets of subjectivity.

While psychoanalysis has elucidated pathological facets of oneness experiences, for example, the characteristic boundary problems of schizophrenics or the way fusion experiences may ward off feelings of disappointment, loss, or Oedipal conflicts (cf. Silverman, Lachmann, & Milich, 1982), it has neglected its adaptive possibilities, particularly what I shall term *non-self-centered subjectivity.* Non-self-centered subjectivity is a psychological–spiritual phenomenon that is implicated in a range of adaptive contexts ranging from psychoanalytic listening to creating or appreciating art to emotional intimacy.[5] It is something many

---

[5]Silverman et al. (1982) illuminate the adaptive as well as the pathological possibilities of oneness experiences, although from a slightly different perspective than I adopt in this book. They emphasize the way that the absorptive union of oneness fantasies can enhance therapeutic success as long as the sense of self is not threatened, while I emphasize the self-expansiveness (and relatedness) of non-self-centered subjectivity. Identity is not "sublated," as Hegel might put it, but extended and enriched. Jones (1995) has helped me articulate this.

people have experienced, e.g., creating art, participating in athletics or religious experiences, or being in love.

It is difficult to paint a picture of such experiences because the English language offers an impoverished vocabulary for evoking non-self-centric states of being. Open to the moment without a sense of time, unselfconscious but acutely aware, highly focused and engaged yet relaxed and without fear, in non-self-centric subjectivity we experience a sense of self-vivification, self-renewal, and self-transformation, and we live, relate, and play with greater creativity, joy, and efficacy than we normally experience. Non-self-centered subjectivity is a source of rejuvenation, sanity, and health. Loewald (1978) depicts some manifestations of these self-enriching facets of non-self-centered subjectivity that nonmeditators, as well as meditators, have probably experienced:

> We get lost in the contemplation of a beautiful scene, or face, or painting, in listening to music, or poetry, or the music of a human voice. We are carried away in the vortex of sexual passion. We become absorbed in ... a deeply stirring play or film, in the beauty of a scientific theory or experiment or of an animal, in the intimate closeness of a personal encounter. (p. 67)

Non-self-centered subjectivity is characterized by heightened attentiveness, focus and clarity, attunement to the other as well as the self, non-self-preoccupied exercise of agency, a sense of unity and timelessness, and non-self-annulling immersion in whatever one is doing in the present. In non-self-centric states of being there is a nonpathological, dedifferentiation of boundaries between self and world; a self-empowering sense of connection between self and world that results in a lack of self-preoccupation and a sense of timelessness, efficacy, and peace.

Moments of non-self-centricity, whether surrendering, merging, yielding, or letting go, seem part of most spiritual traditions. Autonomous, differentiated identity, which is a more self-centric aspect of subjectivity, has been traditionally viewed as the apex of human development by most non-Kohutian psychoanalysts.

Experiences of merger and non-self-centric modes of being have been interpreted by most psychoanalysts, following Freud's unceremonious lead, as symptoms of psychopathology. In the self psychology tradition, for example, fragmentation of self-boundaries and self-cohesion are viewed as symptoms of a vulnerable, besieged, or understructuralized self.[6]

But experiences of self-transcendence, in which there is a loss of self-differentiation and non-self-centricity or a sense of non-self-pre-

---

[6]Atwood and Stolorow (1984) view permeability of self-structure as a sign of a healthy self, provided one also is experiencing self-stability.

occupation, may refer to nonpathological, expanded states of consciousness qualitatively different from more archaic states of nondifferentiation. Although the non-self-centric is usually conflated in psychoanalysis with pathological self-loss, there can be an expansion of self-structure that is not necessarily indicative of an ego defect or a boundary problem and that is self-enriching and not self-annihilating. Because psychoanalysis lacks a psychology of transcendent states, these modes of being do not appear on its map of human development and are assumed to be pathological. Psychoanalysis is guilty of a "prestructural fallacy" (Wilber, 1984) in which these experiences are automatically correlated with archaic development and pathologized.[7]

Neither the psychoanalytic nor the Buddhist view of subjectivity is wrong, but each falls victim to the *pars pro toto* fallacy, reducing subjectivity to the essentialistic core that derives from the unique vantage point they employ in investigating it. Buddhism takes the kaleidoscopic sense of subjectivity derived from its atomistic way of examining self-experience and universalizes it as the essential constituent of subjectivity instead of as an aspect that inevitably emerges when selfhood is examined in the microscopic manner dictated by the Buddhist meditative approach. Psychoanalysis takes the substantive sense of self generated from its telescopic way of investigating subjectivity in analysis and then universalizes that, instead of recognizing it as a facet of subjectivity that necessarily arises when selfhood is examined psychoanalytically. The universalization of the partial psychoanalytic and Buddhist viewpoints impedes the formulation of more encompassing views of subjectivity.

Each conception of subjectivity suggests important correctives to the blindspots of the other. Buddhism's dereified conception of subjectivity can help psychoanalysis in several ways. The experience of meditation practice could teach psychoanalysts that their reified and hypertrophied sense of subjectivity eclipses the fluid "spiritual" aspects of subjectivity: its capacity for non-self-centric manifestations and self-transformations. Through the experience of meditation a psychoanalyst might recognize that the apparently unified narrative that the analysand and the analyst have constructed about the patient's self is, in part, illusory or at least reductionistic. This homogeneous image of the self hides its heterogeneity; the discontinuities, fissures, and anomalous perceptions expressed in the transient and episodic mental states that arise moment after moment. Meditation facilitates, to borrow Foucault's

---

[7]My account has been enriched by Karen Peoples' (1991) unpublished paper entitled "The Paradox of Surrender: Constructing and Transcending the Self."

(1977) terminology in a different but not incompatible context (a Nietzschean-inspired critique of teleological conceptions of history), an "unrealization" (p. 160) of our taken-for-granted, unified identity based teleologically on past experiences and conceptions of ourselves. An unreified and unconstricted sense of subjectivity would be facilitated. Whether or not in the process of doing this Buddhism surreptitiously imposes another conception of self—a non- or antiself—which is also highly conditioned and constricting, is, or ought to be, open to further examination and debate (Joel Kramer, personal communication).

Buddhism could teach psychoanalysts that states of non-self-centricity may represent not a pathological regression or a fragmentation-prone self-structure but spiritual experiences that can enhance self-experience by eroding restrictive identifications and facilitating greater freedom, flexibility, and inclusiveness of self-structures. In detaching one from a hypertrophied, overly self-centric sense of self, meditation practice pursued without adopting the self-denying aspects of Buddhist doctrine opens subjects up to the possibility of greater intimacy; for friendship and love necessitate that we unconstrict and sometimes transcend our normally more restrictive sense of separateness from others and the world. Loosening the grip of excessive self-preoccupation, whether one is deeply immersed in playing a musical instrument, watching an engrossing cultural event, participating in athletics, or making love, often leads to a heightened sense of living.

In psychological and spiritual matters, like in real estate, practically no one voluntarily trades down. The vast majority of meditators would not meditate if they believed they would lose more than they would gain by meditating. Since practically no one—save the Tom Sawyer's of the world—enjoys their own funeral, one wonders what are the unconscious attractions, what are the desires, underlying the Buddhist view of self-nullifying subjectivity? Psychoanalytic understanding of the elaborate and ubiquitous defensive processes or self-protective strategies people employ to ward off pain or maintain self-esteem or self-cohesion can illuminate some of the unconscious attractions of the Buddhist view and increase Buddhist understanding of some of the consequences and dangers of the self-deceptions endemic to Buddhism's stance of self-nullification.

Those who deny their own history (and subjectivity), as Santayana and Freud warned us, are condemned to repeat it. Psychoanalysts could teach Buddhists that in denying their own subjectivity, Buddhists may unwittingly become its prisoner. They may, for example, reenact restrictive childhood interactions such as deferentially submitting to Buddhist teachers cast in the role of idealized authority figures or condemn themselves and experience shame when they fail to garner approval and validation from their idealized teachers.

Disbelieving in the existence of selfhood, as psychoanalysts have discovered in their work with analysands with self-disorders, is appealing to people who feel deep guilt, shame, or fear or wish to erase traumatic history. For example, one way of responding to an abusive childhood or coping with powerful humiliation about a painful experience in one's past is to deny one's existence and thus eliminate any responsibility.

Denying subjective reality can lead to a whitewashing of the amorality of others, which in a given situation might include teachers in spiritual communities. If there is no subject, then there is no impropriety or exploitation. It is all maya, illusion.

The nonexistence of the self might also be attractive to meditators with an apocalyptic sense of the future. For such a person it might be less scary to imagine that one does not exist than to contemplate a self-annihilating nuclear winter in the near future. This is a slight twist on the death-denying–defying "immortality projects" that Ernest Becker (1973) describes in *Denial of Death*. The belief in the nonexistence of selfhood may be a kind of Nietzschean, self-annihilation-for-self-preservation–aggrandizing project. To put it in a quasi-Cartesian framework: I do not exist; therefore I am (not eliminated). Or, to say it in a more modern voice: "When you ain't got nothing," as Bob Dylan said, "you ain't got nothing to lose."

A BIFOCAL VISION OF SUBJECTIVITY

The self is neither an empty signifier signifying nothing as the Buddhists would have it, nor a timeless, transcendental signifier as most psychoanalysts would suggest, but a historical conception that assumes its particular typology and demands to be understood in historical terms.[8] Conceptions of the self are not universal, transcultural, or transhistorical. Different societies construe selfhood in different ways.[9]

[8] I am indebted to David Kastan who suggested this way of putting it in his discussion of kingship in Shakespeare in a chapter, "*Macbeth* and the Name of King" in his forthcoming, *Proud Majesty Made a Subject*. Roland's (1988) attention to cross-cultural differences in child-rearing, socialization, and self-structure is an exception that would justify the generalization about psychoanalysis' tendency to universalize the kind of self-structure that it has highlighted.

[9] Since a particular culture is not a homogeneous unity but the site of internal differences, contradictions, and conflicts, even within the "same" culture selfhood may be viewed in multifarious ways.

Studies of foreign culture demonstrate that conceptions of the self are culturally variable. For example, Bradd Shore's (1985) ethnographic work in Samoa challenges the universal pretensions of the Western humanist conception of an autonomous, Cartesian subject that is concerned with self-knowledge, freedom, and individuation: the "Samoan language has no term corresponding to 'personality, self, character'; instead of our Socratic 'know thyself,' Samoans say 'take care of the relationship'" (p. 65).

Alan Roland's (1988) experiences with non-Western patients in psychoanalysis convinced him of the unconscious "Westerncentric" bias underlying psychoanalytic formulations about human personality. He found through treating Indian patients and supervising the clinical work of Indian and Japanese psychoanalysts and mental health professionals that the psychological makeup of persons in India and Japan is qualitatively different than North European/Americans due to the fundamental differences in the cultural principles of the civilizations and the social patterns and child rearing that these principles shape and delimit. What Roland terms the "familial self," which is rooted in and attuned to the subtle emotional, hierarchical relationships of the family, involving a highly private self engaged in "a constant affective exchange through permeable outer ego boundaries ... [with] high levels of empathy and receptivity to others ... [and an] experiential sense of self ... of a 'we-self' that is felt to be highly relational in different social contexts ..." (pp. 7–8), predominates in Indian and Japanese people and contrasts sharply with the Western individualized self that arises out of contractual, egalitarian relationships geared toward individualism.[10]

The category of selfhood is itself a historic construct. There is evidence that the notion of the self did not exist in the Middle Ages. In feudal Europe, the individual, while a bounded physical entity, had no autonomous sociological status and was conceived as part of a larger network (Dumont, 1970).

Each era, as Levin (1987) argues, produces particular configurations of self. These configurations can be fruitfully understood in terms of literary critic Kenneth Burke's (1941) sense of proverbs as "strategies for dealing with situations" (p. 256). In other words, the specific conceptions of subjectivity generated in a particular era can be viewed, in part,

---

[10]Roland's (1988) characterization of the Western self as essentially individualistic is more true of men than women. Male individualism takes for granted and is indissolubly linked to the support and service of nonautonomous and nonindividualistically oriented women and less privileged men.

as a reflection of and a tactic for dealing with specific historical, psychological, and sociocultural conditions endemic to that age.

As I suggested in Chapter 1, we seem to be living in an age that is
qualitatively different than either fifth-century BC India or nineteenth-
century Europe. The times we confront demand conceptions of subjectivity that recognize the necessity for both groundedness and flexibility
and the capacity for agency and communion. Survival in our world
involves recognizing our embeddedness in lived life and recognizing
selfhood's historicity and relationality (which Buddhism neglects) and
selfhood's non-self-centeredness, self-improvisationality, and self-
transcendence—its spirituality, which could foster connectedness,
care, and enormous flexibility (which psychoanalysis eclipses).

Neither psychoanalysis nor Buddhism provides a complete rendering of subjectivity. Each of their conceptions of subjectivity would be
enriched if the understandings obtained from their different ways of
investigating it were integrated into a more encompassing and inclusive
framework that values their unique insights into the different facets of
subjectivity they each elucidate, while avoiding their limitations, for
example, the nearsightedness of Buddhism and the farsightedness of
psychoanalysis. A more encompassing model of subjectivity would
investigate subjectivity "microscopically" and "telescopically" so that
both its proximate and ultimate facets would emerge. This might help
people better cope with the unique exigencies of late twentieth-century
life.

Subjectivity is composed of various dialectically related facets: self-
centricity and non-self-consciousness, self-assertion and communion,
substantiality and transiency/fluidity, history (embeddedness in time)
and spirituality (the capacity for non-self-centricity/dwelling in timelessness). Psychoanalysis and Buddhism each illuminate and neglect
certain of these facets. Psychoanalysis highlights the self-centricity,
substantiality, and historicity of subjectivity, while Buddhism elucidates its fluid, non-self-centric, and spiritual aspects. Typically, psychoanalysts and Buddhists neglect, if not denigrate, those aspects of subjectivity occluded by their conception of it. One facet of each of the
dialectically connected pairs is thus neglected by each tradition. Buddhists neglect selfhood's substantiality and historicity, while psychoanalysts eclipse its nonsubstantiality and spirituality. To recognize only
one of these aspects is to misperceive the multiplicity of subjectivity.
Neither the self-emptiness and non-self-centricity illuminated by Buddhism nor the self-substantiality and self-centeredness highlighted by
psychoanalysis is, in my view, either primary or superior.

In a well-known and popular formulation, Engler (1986) has claimed that the way psychoanalytic and Buddhist conceptions of self can be reconciled is to realize that both a sense of self and a sense of no-self are necessary for experiencing optimal psychological health and leading a full life. I agree with his claim that the psychoanalytic concern with building a cohesive and integrated self is a "precondition" of the subsequent developmental task of disidentifying from restrictive self-representations and loosening the grip of excessive self-preoccupation. One certainly cannot disidentify from what one is not. But I find his much valorized view within East–West studies that non-self-centeredness is a higher state of being than self-centeredness, that "you have to be somebody before you are nobody" (Engler, 1986, p. 24) is incomplete.

In my view it is incorrect to conceive of non-self-centricity as a higher or final stage of being because a consolidated sense of self and a sense of non-self-centricity are interpenetrating aspects of human experience; alternating positions of being rather than hierarchically ordered stages.[11] The privileging of non-self-centeredness and the consequent devaluation and repression of self-centeredness ultimately engender the egocentric behavior sometimes acted out in Buddhist communities. For as Jung emphasized, when one aspect of subjectivity is consciously overemphasized, its opposite takes on increased unconscious importance.

The expanded conception I am advocating involves thinking more dialectically, granting that both viewpoints have a range of applicability. From my perspective each view of subjectivity is illuminating and myopic. Neither is wrong; they are partial; valid under certain circumstances, false under others. They each eclipse some crucial aspect of subjectivity, and thus offer a partial perspective on it that results in a fragmented portrait of its typology and fate.

The self, in this perspective, might be likened to what Nozick (1989), borrowing from economic theory, terms a "local" as opposed to a "global optimum," by which he means something that is not universally valid but is useful in a particular context. The different stories that psychoanalysis and Buddhism tell about the nature and fate of selfhood are best viewed, in Wittgenstein's (1958) sense, as "reminders for a particular purpose" (I, sec. 127); alternative tools that are useful for different purposes rather than irreconcilable claims.[12]

[11]My position is, in part, an outgrowth of and extensively draws on Ogden's (1991) suggestive interpretation of the notion of "positions" in Melanie Klein.

[12]I do not mean to gloss over the vast differences in assumptions and worldview that separate Buddhism and psychoanalysis. That is beyond the more circumscribed scope of this chapter. I have discussed the differing worldviews underlying each system and their implications elsewhere (cf. Rubin, 1993) and address this issue in greater depth in Chapter 10.

It is impossible to address each of the facets of subjectivity within the scope of this chapter. I shall take one of the dialectical pairs mentioned above and briefly discuss it in terms of the more inclusive perspective I am proposing.

Twentieth-century physics teaches us that we may view phenomena in terms of particles (e.g., things and entities) or waves (e.g., movement and process). When it comes to selfhood most analysts prereflectively adopt the "particulate" perspective: the self as a "fixed entity, separate particle" (Tart, 1990, p. 161). In Buddhist circles the "wave" perspective is valorized.[13]

Life consists of both things and processes, entities and movement. In certain realms of life the "wave" paradigm is useful, while the "particle" paradigm works better for others.

This is illustrated by what might be termed the difference between the "nightwatchman" (Nozick, 1989) and the (Zen) pianist view of the self. The former scrutinizes the world and notices if anything requires attention and the latter participates in his or her craft without undue self-consciousness.

Sometimes when reflecting upon one's conduct, evaluating conflicting moral claims, or planning a course of action, we must fixate the self and view it as a concrete, substantial agent with a history. The concept of a self is necessary as a point of reference for investigating the world, weighing alternatives, attending to feedback, and modifying beliefs or behavior. Decisions and planning cannot occur without a human agent who envisions possibilities, examines alternatives, and chooses specific courses of action. To deny this mode of being is to create chaos and undermine the basis for choice and action.

Yet such self-consciousness would be disastrous for the pianist, who plays best, as Kohut (1984) aptly notes, when she does not notice her fingers. When observing art, participating in athletics, or merging in love, unselfconsciousness, the view of subjectivity as a process, can be valuable. The nightwatchman or particle view of subjectivity interferes with participating in these activates in a complete way. But for people with an insufficiently consolidated self-structure, a wave view of subjectivity would occasion terror and would be absolutely disastrous.

"Most people," as Young (Tart, 1990) observes, "are one-sided, always experiencing self as particle, unfamiliar with self as wave" (p. 161). We often need to be able to oscillate between both modes of being during the course of a day.

---

[13]My bifocal view of subjectivity has greatly benefitted from Shinzen Young's discussion in Tart (1990).

The clinical implications of utilizing the bifocal perspective I have been presenting are at least twofold: Buddhism could teach psychoanalysis about states of dereified, decommodified and non-self-centric subjectivity, while psychoanalysis could teach Buddhists about the recurrent, unconscious organizational patterns and self and object images and associated affects that shape and delimit human life.[14] I will explore this in more detail in Chapter 9.

Taken together, psychoanalysis and Buddhism illuminate a more complete range of the multiplicity that is the self than either pursued alone.[15] A bifocal conception of subjectivity would help us be less myopic about selfhood's important features. Adopting such a complementary viewpoint would enable us to recognize that states of self-centricity and unselfconsciousness are both part of living a full life. The former is necessary to assess situations, formulate plans and goals, and choose among potential courses of action, and the latter is necessary in such experiences as appreciating art, listening to others, participating in athletics, and experiencing love.

A complementary view of subjectivity would help us avoid Buddhism's farsightedness and psychoanalysis' nearsightedness. We would not eclipse the view of self-fluidity, non-self-centricity and self-transcendence suggested by Buddhism; nor would we deemphasize the sense of self-substantiality, historicity, and agency recognized by psychoanalysis.

REFERENCES

Atwood, G., & Stolorow, R. (1984). *Structures of subjectivity: Explorations in psychoanalytic phenomenology.* Hillsdale, NJ: Analytic Press.
Becker, E. (1973). *Denial of death.* New York: Free Press.
Borges, J.L. (1970). *Labyrinths: Selected stories and other writings.* New York: New Directions.

[14]Buddhists may find my approach toward Buddhism both too conservative *and* too radical. In terms of the former, I do not jettison, and even attempt to salvage, subjectivity; and in terms of the latter because I break from the almost universal tendency of valorizing the Buddhist perspective and question its status as received Truth rather than as a contingent, historical creation that both illuminates and obscures certain aspects of human life. I take solace in the opening remark of the Dalai Lama, the spiritual head of Tibetan Buddhism, to the organizers of a recent conference on Buddhism and Western science: "Remember, Buddhism is not everything" (Lobsang Rapgay, personal communication).

[15]My rendering of subjectivity is necessarily incomplete. Investigators utilizing different theoretical templates and investigative tools will detect other dimensions of selfhood.

Boucher, S. (1988). *Turning the wheel: American women creating the new Buddhism.* San Francisco: Harper & Row.

Burke, K. (1941). *The philosophy of literary form: Studies in symbolic action.* New York: Vintage Books.

Dumont, L. (1970). *Homo hierarchicus: The caste system and its implications.* Chicago: University of Chicago Press.

Engler, J. (1984). Therapeutic aims in psychotherapy and meditation: Developmental stages in the representation of self. In K. Wilber, J. Engler, & D. Brown (Eds.), *Transformations of consciousness: Conventional and contemplative perspectives on human development* (pp. 17–51). Boston: Shambhala.

Epstein, M. (1989). Forms of emptiness: Psychodynamic, meditative and clinical perspectives. *Journal of Transpersonal Psychology, 21*(1), 61–71.

Epstein, M. (1990). Psychodynamics of meditation: Pitfalls on the spiritual path. *Journal of Transpersonal Psychology, 22*(1), 17–34.

Epstein, M. (1995). *Thoughts without a thinker: Psychotherapy from a Buddhist perspective.* New York: Basic Books.

Foucault, M. (1977). Nietzsche, genealogy, history. In D. Bouchard (Ed.), *Language, counter-memory, practice* (pp. 139–164). Ithaca, NY: Cornell University Press.

Gill, M. (1982). *Analysis of transference,* Vol. 1. New York: International Universities Press.

Gold, J. (1992). An integrative-systemic approach to severe psychopathology in children and adolescence. *Journal of Integrative and Eclectic Psychotherapy, 11,* 55–70.

James, W. (1890). *The principles of psychology.* New York: Dover.

Jones, J. (1995). *The real is the relational: Psychoanalysis as a model of human understanding.* Unpublished manuscript.

Kohut, H. (1984). *How does analysis cure?* Chicago: University of Chicago Press.

Kramer, J., & Alstad, D. (1993). *The guru papers.* Berkeley, CA: North Atlantic Books.

Laing, R.D. (1965). *The divided self.* New York: Penguin.

Langs, R. (1976). *The bipersonal field.* New York: Jason Aronson.

Levin, D.M. (1987). Introduction. In D.L. Levin (Ed.), *Pathologies of the modern self: Postmodern studies on narcissism, schizophrenia, and depression* (pp. 1–20). New York: New York University Press.

Loewald, H. (1978). *Psychoanalysis and the history of the individual.* New Haven, CT: Yale University Press.

Meyers, H. (1989). Introduction. In A. Cooper, O. Kernberg, & E. Person (Eds.), *Psychoanalysis: Towards the second century* (pp. 135–142). New Haven, CT: Yale University Press.

Mitchell, S.A. (1991). Contemporary perspectives on self: Toward an integration. *Psychoanalytic Dialogues: A Journal of Relational Perspectives, 1*(2), 121–172.

Mitchell, S.A. (1983). *Relational concepts in psychoanalysis.* Cambridge, MA: Harvard University Press.

Modell, A.H. (1985). The two contexts of the self. *Contemporary Psychoanalysis, 21*(1), 70–90.

Nicholson, W. (1989). What the ultimate "observation method" overlooks. Unpublished paper.

Nozick, R. (1989). *The examined life: Philosophical meditations.* New York: Simon & Schuster.

Nyanaponika, T. (1973). *The power of mindfulness.* San Francisco: Unity Press.

Ogden, T.H. (1991). An interview with Thomas Ogden. *Psychoanalytic Dialogues: A Journal of Relational Perspectives, 1*(3), 361–376.

Peoples, K. (1991). The paradox of surrender: Constructing and transcending the self. Unpublished paper.

Roland, A. (1988). *In search of self in India and Japan: Toward a cross-cultural psychology.* Princeton, NJ: Princeton University Press.

Rubin, J.B. (1993). Psychoanalysis and Buddhism: Toward an integration. In G. Stricker & J. Gold (Eds.), *Comprehensive textbook of psychotherapy integration* (pp. 249–266). New York: Plenum Press.

Rubin, J.B. (submitted). *The blindness of the seeing I: Perils and possibilities in psycho-analysis.* New York: New York University Press.

Schafer, R. (1989). Narratives of the self. In A. Cooper, O. Kernberg, & E. Person (Eds.), *Psychoanalysis: Towards the second century* (pp. 153–167). New Haven, CT: Yale University Press.

Schuman, M. (1991). The problem of self in psychoanalysis: Lessons from Eastern philoso-phy. *Psychoanalysis and Contemporary Thought, 14*(4), 595–624.

Shore, B. (1985). *Sala'ilua: A Samoan mystery.* New York: Columbia University Press.

Silverman, L., Lachmann, F., & Milich, R. (1982). *The search for oneness.* New York: International Universities Press.

Stolorow, R., & Lachmann, F. (1984/5). Transference: The future of an illusion. *The Annual of Psychoanalysis, 12/13*, 19–37.

Sullivan, H.S. (1940). *Conceptions of modern psychiatry.* New York: Norton.

Sullivan, H.S. (1950). The illusion of personal individuality. In H.S. Sullivan (Ed.), *The Fusion of Psychiatry and Social Science* (pp. 198–226). New York: Norton.

Tart, C. (1990). Adapting Eastern spiritual teachings to Western culture: A discussion with Shinzen Young. *Journal of Transpersonal Psychology, 22*(2), 149–165.

van de Wetering, J. (1975). *A glimpse of nothingness: Experiences in an American Zen community.* Boston, MA: Houghton Mifflin.

Walsh, R. (1989). Personal communication.

Walzer, M. (1988). *The company of critics: Social criticism and political commitment in the twentieth century.* New York: Basic Books.

Wilber, K. (1984). The developmental spectrum and psychopathology; Part I, Stages and types of pathology. *Journal of Transpersonal Psychology, 16*(1), 75–118.

Wittgenstein, L. (1958). *Philosophical investigations.* New York: MacMillan.

Wollheim, R. (1984). *The thread of life.* Cambridge, MA: Harvard University Press.

Wood, M. (1989). Montaigne and the mirror of example. *Philosophy and Literature, 13*(1), 1–14.

II
Health and Illness

4

The Emperor of Enlightenment
May Have No Clothes

A Zen master's life is one continuous mistake.
—Dogen

Psychoanalysis and Buddhism are both deeply concerned with the problem of human suffering. One key purpose of psychoanalysis, according to Freud, was to eliminate neurotic misery. A central task of Buddhism is to achieve Enlightenment, which is said to eradicate suffering. Each tradition presents a highly sophisticated theory and methodology for attempting to alleviate human misery. There are strengths and limitations in each conception.

In this chapter, I shall challenge certain foundational assumptions of the Theravadin Buddhist conception of Enlightenment in light of psychoanalytic conceptions of self and cure, and I shall utilize Buddhist perspectives on selfhood in order to pinpoint certain limitations in psychoanalytic conceptions of health.

Since Buddhism, like psychoanalysis, is heterogeneous, composed of various schools of thought adopting different philosophies and types of practices, there is thus no monolithic definition of Enlightenment. It has been described in various ways in different Buddhist traditions and even within the "same" tradition. To Zen master Dogen, the founder of Soto Zen, Enlightenment meant "intimacy with all things," while an esteemed Tibetan Buddhist monk–psychiatrist has described it as "no unconsciousness" (Lobsang Rapgay, personal communication). In the Theravadin Buddhist model presented in this chapter, Enlightenment is described as completely purifying the mind of "defilements," e.g., greed, hatred, and delusion, which is said to result in the total cessation of suffering.

I shall raise two sets of questions about this conception: its possibility and its desirability. My remarks are meant in the spirit of encouraging further dialogue about important, complex, and neglected issues.

In an oft-quoted remark, Freud maintained that one purpose of analysis was to replace "neurotic misery" with "common human unhappiness." From the psychoanalytic perspective, some suffering—the neurotic kind—can and should be ameliorated. But *some* suffering is basic to human existence. Buddhism differs from psychoanalysis in its belief that suffering itself can be eliminated when one attains Enlightenment.

Enlightenment is often presented as the summum bonum or highest good of human existence. Buddhists, as well as representatives of other Eastern traditions, caution that descriptions of Enlightenment are inadequate. Nozick's (1989) description is worth quoting. Enlightenment is said to be

> blissful, infinite, without boundaries or limit, ecstatic, full of energy, pure, shining, and extremely powerful ... it feels like an experience of something, an experience revelatory of the nature of a deeper reality.... This experience seems to reveal reality to be very different from the way it ordinarily appears. (pp. 243–244)

The self, as well as reality, is then experienced very differently with its boundaries either "extended or dissolved" (p. 246):

> Not only does the person feel during the enlightenment experience that his deepest self is very different, often he is transformed as a result of the experience. The enlightenment experience of a very different mode of self-organization enables him also to encounter the everyday world differently, now less clouded or distorted by the interests of the limited self. (p. 247) ... The enlightenment experience not only ends your identifying with the self as a particular delimited entity, it might be an experience of being no entity at all.... You don't have to possess or choose any essence at all, then to think you have one is a mistake. (p. 248)

Enlightened is defined in the most comprehensive study of enlightened meditators that has ever been done, as the permanent transformation of consciousness leading to freedom from suffering and life without discontent: a state of "perfect wisdom and compassion and freedom from any kind of suffering" (Brown & Engler, 1986b, p. 207).

In Theravadin Buddhist psychology, the Arahant, one worthy of praise for slaying the demons of greed, hatred, and delusion, is the exemplar of ideal mental health and Enlightenment. The Arahant is described as a being in whom "no unhealthy mental factors whatsoever arise in the mind' (Goleman, 1980, pp. 133–134). The Arahant evidences an

1. absence of greed for sense desires, anxiety, resentments, or fears of any sort; dogmatisms such as the belief that this or that is "the Truth"; aversions to conditions such as loss, disgrace, pain, or blame; feelings of lust or anger; experience of suffering; need for approval, pleasure, or praise; desire for anything for oneself beyond essential and necessary items

2. prevalence of impartiality toward others and equanimity in all circumstances; ongoing alertness and calm delight in experience, no matter how ordinary or even boring; strong feelings of compassion and loving kindness; quick and accurate perception; composure and skill in taking action. (Goleman, 1980, p. 134)

Enlightenment, that condition of panoramic awareness, nonreactivity, selflessness, wisdom, compassion, and love that has been deified for thousands of years in various religious traditions, appears to be an ideal that is completely innocent of self-deception, corruption, and suffering. It promises to eliminate the egoism, desire, and fear that most people feel mired in (Kramer & Alstad, 1993).

Enlightenment is a compelling vision of what we might become. A world beyond suffering is not to be denied out of hand or sloughed off easily. In assuming that human suffering is inevitable, analysis adopts a depressogenic stance toward the universe, which itself contributes to human suffering. The analyst in me who is keenly aware of the limitations of this vision of the world longs for a viable alternative, and the contemplative in me believes that the Buddhist vision may be the most interesting candidate we have.

An American once asked the Dalai Lama how one could identify true Buddhist teachers. The Dalai Lama replied, "Watch them. See how they behave" (Tworkov, 1994, p. 157). The rash of grossly self-centered and conspicuously unenlightened behavior exhibited by Buddhist teachers in recent years—documented by Sandy Boucher (1988) in *Turning the Wheel: American Women Creating the New Buddhism*—challenges the pervasive idealization of the ideal of Enlightenment and suggests that it would be prudent both to not uncritically accept this notion and to inquire further into the psychological dynamics that might be unconsciously operating.[1] In order to do this it will be helpful to briefly explicate the meditative process as it relates to Enlightenment.

In his "deconstructivist" model of the normative Theravadin meditative process Engler (n.d.) has likened the basic task of meditation to a deconstruction of the nonveridical perceptual process and the "constructions which bias perception and thought" (p. 725) and create a "world" of internal representations of self and others. Meditation practice, in this model, "traces the cognitive and affective pathways by which self and objects literally come into being ... [and one] learns to correct the faulty reality testing which leads to non-veridical percepts of self and objects" (p. 724).

As the meditator's distractedness lessens and his concentration and attentiveness deepens, mental phenomena are recognized immediately after their arising. "Dispelling the illusion of compactness" (Nyanamoli, 1976) occurs when the meditator's concentration and attentiveness is sufficiently refined and their capacity to remain steadily aware for extended periods of times increases. Engler (1986) provides a cogent summary of the series of transformations that occur:

> The normal sense that I am a fixed, continuous point of observation from which I regard now this object, now that, is dispelled ... my sense of being a separate observer or experiencer behind observation or experience is revealed to be the result of a perceptual illusion, of my not being normally able to perceive a more microscopic level of events. When my attention is sufficiently refined through training ... all that is actually apparent to me from moment to moment is a mental or physical event and an awareness of that event.... No enduring or substantial entity or observer or experiencer or agent—no self—can be found behind or apart from these moment-to-moment events to which they could be attributed (an-atta = no-self). (p. 41)

When the meditator's attentiveness remains stabilized at this level of perception, deeper insights about the nature of self and object representations emerge: The meditator experiences

---

[1]Psychoanalysis and psychoanalysts are not without scandals and exploitation as the Jung–Sabina Spielrein affair demonstrates. The absence of data about such incidents in psychoanalysis make it more difficult to examine. Because it is not well-documented, it is difficult to compare it with acting out in spiritual communities. Since this topic is tangential to the main focus of the terrain I will be exploring in this chapter and will take me too far afield to explore the multiple meanings and functions of this sort of behavior, I will not discuss it here. Questions for future researchers might include such things as: Are therapeutic scandals less frequent because they are more suppressed or because they occur less often? If they occur less often, is it because of the explicit emphasis in psychoanalytic theory and training on the analysand's transference and the analyst's countertransference? and so forth.

how a self-representation is constructed in each moment as a result of an interaction with an object ... and conversely, how an object appears not in itself ... but always relative to my state of observation.... I discover that there are actually no enduring entities or schemas at all; only momentary constructions are taking place. (Engler, 1986, p. 42)

As nonselective and reactive attentiveness continues to deepen, the meditator experiences that

the stream of consciousness literally break[s] up into a series of discrete events which are discontinuous in space and time.... Representation and reality construction are therefore discovered to be discontinuous processes.... When this total moment-to-moment "coming to be and passing away" (udayabbaya) is experienced, there is a profound understanding of the radical impermanence (anicca) of all events.... I become aware of the selflessness (anatta) of mind, body, external objects and internal representations ... any attempt to constellate enduring self and object representations, or to preferentially identify with some self-representations as "me" and expel ... or repress ... others as "not-me," is experienced as an equally futile attempt to interrupt, undo or alter self and object representations as a flow of moment-to-moment constructions. (Engler, 1986, pp. 42–44)

As the meditator sees through and lets go of her illusory internal world of representations, she experiences a profound loss: the ontological foundations of her world are undermined. This initiates a "mourning" process involving pain and despair over the impossibility of recovering the lost object (world). This leads to what Engler (n.d.) terms a "reorganization" of one's inner world involving

decathecting both (a) the need for the lost external object and (b) its internalized image. (p. 728) ... Unlike normal mourning ... the meditator is not simply confronted with the loss of a single object but with the loss of all his objects, of the object world as such, internal as well as external. (p. 730) ... In meditation, there is thoroughgoing object loss. (p. 731) ... [I]n meditation there is a renunciation of self-object ties altogether.... There is no new identification and no more object seeking.... In meditation all object ties are finally "outgrown." (p. 733)

This mourning of self and object loss and the renunciation of all object ties may be viewed, according to Engler (n.d.), "as the final step in separation-individuation: the end of suffering and ultimate individuation" (p. 734).

The concept of Enlightenment, as I understand it, is underwritten by at least two assumptions about human life and development. It presupposes that the mind can be permanently transformed and completely purified and that there is a stage of development, beyond egocentricity and the vicissitudes of conflict, that is irreversible and everlasting.

Spiritual teachers are often presented as being beyond self-blindness. There seems to be little evidence that many spiritual teachers discourage such idealized images. This is not surprising, since admitting that one had pockets of self-blindness or proclaiming one's psychological fallibility would not exactly attract a steady following of students.

In 1983, five of the six most esteemed Zen Buddhist masters in the United States, who presumably were selected by an enlightened teacher abroad to teach, were involved in grossly self-centered and conspicuously unenlightened behavior, such as sexually exploiting nonconsenting students and illegally expropriating funds from the community.[2] Leaving aside the fact that this is a complex phenomenon that is based on multiple historical, cultural, and psychological factors, some of which I shall explore in more detail in Chapter 9, an urgent question confronts us: If enlightenment is an irreversible transformation and purification of the mind, then why have these troubling incidents occurred?

Explanations of these incidents have been unconvincing. The problem tends to be (1) rationalized as, in the words of one Buddhist teacher, "crazy wisdom" that the unenlightened cannot understand; (2) minimized as an isolated aberration of an unevolved individual; or (3) connected to sociocultural differences between Asian and Western cultures.

Lest one defensively claim that these problems are incidental or accidental, it is worth reflecting on one of the conclusions of the previously cited study of enlightened Buddhist meditators that these individuals evidenced residues of psychopathology as well as extraordinary development, remarkable clarity, and profound compassion (Brown & Engler, 1986b). These meditators were, in the words of Drs. Brown and

---

[2]It would not surprise me if this troubling doubt were dealt with by dismissing it with the ready-made conclusion that not all Buddhist teachers are enlightened so that any disturbing incidents of egocentricity and exploitation on the part of a Buddhist teacher do not tarnish the ideal of Enlightenment. This strikes me as both a misguided and irresponsible response to incidents that should be explored more extensively. A teacher is presumably one who through discipleship (with an enlightened master?) and extensive work on him- or herself has attained significant self-understanding and self-realization. If those sanctioned to teach by indigenous practitioners who are often viewed as Buddhist "masters" do not evidence the highest possibilities for psychological and spiritual health that are theoretically posited, then one wonders who does illustrate these claims. Even if the questions I am attempting to raise were challenged on the grounds that the status and validity of Enlightenment remain unaffected by these incidents of self-blindness and are, in fact, irrelevant to them, I think it would be a shame to ignore or rationalize this conduct rather than to utilize it to explore a certain unconsciousness in the hallowed and essentially unquestioned Buddhist ideal.

Engler, "not without conflict, in a clinical sense. They show evidence for the experience of drive states and conflictual themes such as fears [and] dependency struggles" (pp. 210–211). Brown and Engler's (1986a) Rorschach studies of enlightened meditators indicate that "these allegedly enlightened advanced practitioners are not without intrapsychic conflict ... each of these Rorschach's evidenced idiosyncratic conflictual themes such as fear of rejection; struggles with dependency and needs for nurturance; fear and doubt regarding heterosexual relationships; fear of destructiveness" (pp. 188–189).

The observations of Jack Kornfield (1988), an esteemed American teacher of Theravadin Buddhism who studied with indigenous Buddhist masters in Asian monasteries for many years and has taught Buddhist meditation throughout the world for almost two decades, concur. He points to two of the "limitations" and insufficiencies of meditation practice: (1) unresolved personal, relational, and occupational issues that meditation practice does not alter in meditation students and teachers, and (2) "major upheavals and problems around power, sex, honesty, intoxicants etcetera ... in a majority of the twenty or more largest centers of Zen, Tibetan, and Vipassana practice in America ... centering on the teachers (both Asian and American) themselves" (p. 10).

Brown and Engler's findings were explained in terms of the failure of the subjects of their study to attain the highest "stages" of Enlightenment. It was proposed that these subjects had only attained lower levels of Enlightenment and that conflict might be eradicated on higher stages. This claim was neither tested nor proven, so it remains essentially speculative in the absence of further evidence.

The psychological reductionism and oversimplicity of linear stage models of mind obscures the complexity and asymmetry of human development and mental life; for most people there probably is no uniformity to their identity and stage of development. One's empathy for others, for example, may be significantly less developed than one's own self-knowledge. The vast knowledge gained by psychoanalysts about the ubiquity of self-unconsciousness casts grave doubt on Buddhist claims about permanent and irreversible self-transformation and wisdom.

The practice of psychoanalysis teaches one that mental life is fluid rather than static, involving the continual, dialectical interplay of various states of consciousness, subject-positions, or self-states and modes of being that are sometimes at cross-purposes and in conflict. Living a human life is thus more like sailing, confronting exigencies that are both everchanging and unpredictable, than attaining any sort of permanent and irreversible state. Conflict can no more be eliminated from mental life in the psychoanalytic model than the vagaries of the wind can be permanently eliminated from sailing.

Buddhist models of the mind also acknowledge that the mind, like the universe, is always in flux. But, with its recognition that everything changes—except Enlightenment, which is posited as an unchanging achievement—Buddhism attempts to eat its cake of flux and have it too.

Understanding the mind, like achieving physical health, is not a static attainment. Mental health, as psychoanalyst Melanie Klein (1960) suggested, is an ongoing job requiring continual attentiveness and persistence. Freedom, as John Meynard Keynes aptly noted, requires eternal vigilance (Joel Kramer, personal communication).

The acting out in spiritual communities on the part of Buddhist teachers, as well as the previously cited empirical research on Enlightenment, suggests that psychological conditioning from the past that inevitably warps personality cannot be completely eradicated and that there is no conflict-free stage of human life in which the mind is permanently purified of conflict. This is consistent with psychoanalytic insights about the essential nontransparency of the human mind; that is, the inevitability of unconsciousness and self-deception.

For an individual to be enlightened, they would have to be certain that they were completely awake without any trace of unconsciousness or delusion. Even if that existed in the present, it is not clear to me how one could know for certain that would never change in the future (cf. Kramer & Alstad, 1993). From the psychoanalytic perspective, a static, conflict-free sphere—a psychological "safehouse"—beyond the vicissitudes of conflict and conditioning where mind is immune to various aspects of affective life such as self-interest, egocentricity, fear, lust, greed, and suffering is quixotic. Since conflict and suffering seem to be inevitable aspects of human life, the ideal of Enlightenment may be asymptotic, that is, an unreachable ideal.

In questioning the ideal of Enlightenment and claiming that the Emperor of Enlightenment, if you will, may have no clothes, I do not mean to neglect or devalue those moments of extraordinary clarity, self-acceptance, inner spaciousness, peace, and abiding love that meditators for millennia have experienced. Nor do I wish to deny that Buddhist teachers may have deep insights about self and life that might be illuminating for others as long as the context of teaching is based on an egalitarian, mutually respectful and empowering relationship rather than an authoritarian, hierarchical connection characterized by deification and submission.

What I do wish to question is the notion of a mind that is somehow permanently without any ripple of unconsciousness, self-deception, and selfishness. In doing this I hope that it is still clear that I recognize and have deep respect for the profound transformative possibilities of meditative *practice*, which teaches psychoanalysis that humans are capable of much greater self-awareness, compassion, and inner peace then psychoanalysts usually recognize.

While Buddhist practice may be profoundly self-enhancing, Buddhist ideals of Enlightenment are not without certain problematic consequences, which become more apparent when juxtaposed with a particular psychoanalytic conception of health, namely, self-integration or self-enrichment (cf. Atwood & Stolorow, 1984; Rizzuto, 1994). By self-integration I refer to the experience that patient analysands in successful treatment have of being more able to know, tolerate, embrace, integrate, and communicate formerly disavowed facets of their inner, personal reality. This fosters greater self-knowledge and self-acceptance and leads to an enriched engagement with oneself and the world. One's perspective of self and the world becomes less rigid and more inclusive as new internal and external experiences are accommodated to more readily. One becomes more tolerant of difference, in oneself and others, and develops a less egocentric and more inclusive and compassionate perspective on self and others. This perspective on health suggests that the Buddhist view is not only incomplete, it is, in at least three ways, also limiting.

First, eradicating suffering is central to the Buddhist view of health. The one thing I teach, asserted Buddha, is the cessation of suffering. A deeply unconscious assumption in Buddhism is that suffering is bad. Is all suffering bad? Certainly suffering is painful. It is natural to wish to avoid it, but if we reflect on times of self-transformation and growth in our own lives, they were probably often accompanied by suffering. Suffering that is worked through can deepen and enrich a human life, by generating greater knowledge, openness, sensitivity, compassion, and passion.

Because suffering can be edifying, one wonders why Buddhism is so preoccupied with the goal of removing it. Does the Buddhist goal of eradicating suffering bespeak an unconscious aversion to life that could actually be self-limiting because it removes one from engagement with life's existential and emotional vicissitudes and the self-knowledge that it can foster? In a world in which the suffering of women or Afro-Americans was prematurely removed, the discontent that sowed the seeds of the feminist and civil rights movements may have been uprooted and the moral outrage that fueled constructive social change and transformation may have been compromised.

The ideal of Enlightenment can foster self-impoverishment by en-

couraging meditators to unconsciously renounce and become detached
from the complexity and passion of an embodied human existence.
Might it not be an enriched Buddhism that simultaneously worked on
eradicating human misery while it investigated the possible unconscious meanings and benefit of its own unconscious attachment to the
project of disengaging from this facet of life?

Second, in emphasizing the deconstruction of self, the decathexis
of both internal representations of other people and ourselves, the emp-
tying of our inner psychological world and the mourning that it fosters,
the Theravadin Buddhist model delineated earlier in this chapter pin-
points one important facet of the healing process. Nonattachment, in the
Buddhist sense, to outmoded ties to self and others certainly contributes
to the process of lessening human suffering and expanding the possibilities for human liberation. As attachment and the grasping and aversion it fosters lessens, we are then freer to respond with greater openness
and concern to ourselves and others.

Is the emptied inner world of Enlightenment that Engler depicts
spacious and free or self-alienating and impoverished? Are all ties with
others negative? Are all relationships enervating? Do relationships ever
promote empathy and the mitigation of suffering? Was the novelist Toni
Morrison (1993) way off the mark when she said in a recent interview
that love is a space where freedom can be negotiated (p. 113)?

Intimacy can be promoted or obstructed by Buddhism as well as
psychoanalysis. Buddhist theory can simultaneously encourage and
inhibit greater human relatedness. Its commitment to a non-self-absorbed
relationship with self, others, and the world can contribute to less
exploitative conduct, while its self-nullifying view of subjectivity may
interfere (as I suggested in Chapter 3) with genuine intimacy that is
based, at least in part, on mutuality between two individuated individuals. Individuation does not flourish on the soil of self-negation.

The Theravadin Buddhist stance toward the world as embodied in
the ideal of Enlightenment neglects the radical insight in psychoanalyst
Heinz Kohut's (1977) claim about the lifelong necessity of vital ties with
other human beings. It seems to minimize our inevitable embeddedness
in relationships and their potential value. Participation in relationships
can contribute to both greater self-awareness and increased self-validation.
An enhanced sense of ourselves and capacity for agency and moral
action can result.

Like the philosopher Immanuel Kant's (1968) dove, which regarded
the resistance of the air as an obstacle and imagines that it could fly
better in the vacuum of "empty space" (p. 47), Buddhism seems to view
affective life as an obstacle to living instead of an irreplaceable aspect of

life. In the model of self-integration or enrichment that I am advocating, past experiences and affective life, including relations with self and others, are not so much weeds to be eliminated as they are manure to fertilize.

In such a conception, the self can be likened to a symphony composed of a variety of instruments—consciousness and unconsciousness, self-centeredness and selflessness, rationality and imagination, and so forth—each with its own idiosyncratic sound and application. The self is impoverished if certain instruments are not played. The best music occurs when no instrument dominates or is excluded and when there is communication and cross-fertilization between them.

Third, the self-deconstruction that meditative practice promotes can lead to irresponsible self-disengagement[3] as well as self-impoverishment. Assuming that self and object representations are merely, in Engler's words, "constructs" (artifacts of the constructivist activity of perception) fosters a denial or minimization of self-existence that creates greater self-unconsciousness. For we are then predisposed to not notice facets of ourselves, such as greed or self-centeredness, that clash with our cherished self-conceptions. The privileging of self-negation and the consequent devaluing and repression of self-centeredness ultimately engenders the egotistical behavior sometimes acted out in Buddhist communities. The destructive self-centeredness that has plagued certain spiritual teachers and communities and caused suffering for all concerned is the natural result of such a self-nullifying stance toward human subjectivity. The perceptual disavowal of self is not the experiential working through of its historical legacy. The troubling interactions and incidents of exploitation that I have described above are sown from the seeds of such self-disavowal.

Self-disavowal can be particularly disastrous for people who have traditionally been marginalized in Western society such as women and racial or religious minorities. Such people have often been crushed or invalidated on a sociocultural as well as personal level. In theoretically calling into question the very category of persons, deconstructivist theories about the self interfere with a viable view of human agency, and thus inhibit political engagement. For if there is no subject, then there is no

[3]The recent movement known as "engaged Buddhism," which utilizes Buddhist perspectives to address social issues such as ecology and the peace movement, actually illustrates rather than challenges my claim that Buddhism can lead to a restrictive detachment from the inner and outer world. For you only need to engage what was formerly disengaged. Engaged Buddhism is only necessary because Buddhism has previously fostered disengagement. Also, Buddhists can and often do engage the outer world in order to fight for social justice only to unconsciously detach from their own inner world of thoughts, feelings, and fantasies.

one who is alienated or oppressed, and thus no evil to challenge and no one to contest it. Such a stance toward the self denies and minimizes one's oppressed position within the asymmetrical social status quo. This can be crippling or disabling by implicitly or explicitly perpetuating the self-alienation and marginalized status of the oppressed even as it may theoretically undermine authoritarian doctrines that present reified visions of persons and are oppressive.

\The emptied inner world of the Enlightenment experience sounds more barren and impoverished than expansive and liberating. Renouncing self and other leads not to freedom but to self-alienation. Freedom derives not from renunciation of ties to self and others but from freedom within the context of relatedness\ In a world like ours in which profound self-alienation and emotional disconnection predominate, the psychoanalytically derived view of self-enrichment I have presented as an alternative to the Buddhist model of self-renunciation, a view of humans that values self-expansion and enrichment and connectedness to others, might foster less suffering than a model of self-renunciation and self-purification.\4 In Chapter 10, I will discuss this issue.

\The model of self-enrichment is a valuable way of thinking about the analytic process and health. A great deal of human suffering might be alleviated if more people, non-Buddhist and Buddhist alike, engaged in such a process. \

\But Buddhism teaches psychoanalysis that the notion of self-integration can also be a limiting and imprisoning type of self-experience, a suboptimal state of being that may foster self-restriction and self-alienation and thereby contributes to human suffering\ Since selfhood is not a singularly definable entity but a heterogeneous and complex phenomenon that is context-dependent, singular notions like the integrated self may miniaturize and subjugate selfhood's possibilities (cf. Hillman, 1975) by obscuring and limiting its multidimensionality. Facets of self-experience that do not fit into preexisting images of who one really is are neglected or not assimilated. This impedes hospitality toward facets of ourselves that do not fit into the unified narrative we have constructed about ourselves. Opportunities for complicating and transforming one's sense of self are thus severely limited (cf. Sennett, 1970). The always unfinished and open nature of the self, in the view I am briefly sketching, is an achievement to be further cultivated rather than a defect to be remedied.[5] The notion of an integrated self may thus constitute what

[4]The renunciation model can also be found in Freudian analysis.

[5]Psychoanalyst Adam Phillips' suggestive remarks on the value of the incompleteness and openness of self at a conference on psychoanalysis and Buddhism in New York City in April 1994, "Healing the Suffering Self: A Dialogue among Psychoanalysts and Buddhists," spurred me to pursue this line of thinking.

Erich Fromm (1941) might term an "escape from freedom": the ever-renewing possibility and responsibility for creating who we are.

Self-integration and self-deconstruction may both be part of what is necessary to achieve psychological health and reduce psychological suffering. I prefer a world in which we neither assume the self's sovereignty nor absolutize its provisionality. If Buddhism has correctly pinpointed the dangers of absolutizing the self, then it may now be time for Buddhists to recognize, perhaps with the help of psychoanalysis, that there is a hidden and pernicious cost to absolutizing its view of the fictionality of the self. To fail to do so also brings suffering, as the incidents in spiritual communities so vividly demonstrate.

In a world in which self-investment and empathy and care for others were seen as complementary—two interpenetrating facets of what it meant to be a human being—we might experience that liberating bifocal perspective that the poet W.H. Auden (1969) conveyed when he wrote: "cosmic trivia/we all are, but none of us are unessential" (p. 40).

REFERENCES

Atwood, G., & Stolorow, R. (1984). *Structures of subjectivity: Explorations in psychoanalytic phenomenology*. Hillsdale, NJ: Lawrence Erlbaum.

Auden, W.H. (1969). *Epistles to a godson and other poems*. New York: Random House.

Boucher, S. (1988). *Turning the wheel: American women creating the new Buddhism*. San Francisco: Harper & Row.

Brown, D., & Engler, J. (1986a). The stages of mindfulness meditation: A validation study. Part I: Study and results. In K. Wilber, J. Engler, & D. Brown (Eds.), *Transformation of consciousness: Conventional and contemplative perspectives on human development* (pp. 161–191). Boston: Shambhala.

Brown, D., & Engler, J. (1986b). The stages of mindfulness meditation: A validation study. Part II: Discussion. In K. Wilber, J. Engler, & D. Brown (Eds.), *Transformation of consciousness: Conventional and contemplative perspectives on human development* (pp. 191–217). Boston: Shambhala.

Engler, J. (n.d.). The practice of insight. Unpublished manuscript.

Engler, J. (1984). Therapeutic aims in psychotherapy and meditation: Developmental stages in the representation of self. In K. Wilber, J. Engler, & D. Brown (Eds.), *Transformations of consciousness: Conventional and contemplative perspectives on development* (pp. 17–51). Boston: Shambhala, 1986.

Goleman, D. (1980). Mental health in classical Buddhist psychology. In R. Walsh & F. Vaughan (Eds.), *Beyond ego: Transpersonal dimensions in psychology* (pp. 131–134). Los Angeles, CA: Tarcher.

Hillman, J. (1975). *Re-visioning psychology*. New York: Harper & Row.

Kant, I. (1968). *Critique of pure reason*. New York: St. Martin's Press.

Klein, M. (1960). On mental health. In M. Klein, *Envy and gratitude and other works (1946–1963)* (pp. 268–274). New York: Delta.

Kohut, H. (1977). *The restoration of the self*. New York: International Universities Press.

Kornfield, J. (1988, Summer). Meditation and psychotherapy: A plea for integration. *Inquiring Mind*, 10–11.

Kramer, J., & Alstad, D. (1993). *The guru papers: Masks of authoritarian power*. Berkeley, CA: North Atlantic Books.

Morrison, T. (1993). The art of fiction. *The Paris Review, 128*, 83–125.

Nozick, R. (1989). *The examined life: Philosophical meditations*. New York: Simon & Schuster.

Nyanamoli, B. (Trans.). (1976). *The path of purification*. Boulder, CO: Shambhala.

Rizzuto, A.M. (1994). *Sound and sense: Words and the paradox of the suffering person*. Unpublished manuscript.

Sennett, R. (1970). *The uses of disorder: Personal identity and city life*. New York: Norton.

Tworkov, H. (1994). *Zen in America*. New York: Kodansha International.

5

Psychoanalytic Treatment with a Buddhist Meditator

The case study has a long and distinguished history within psycho-analysis, being the principal approach of the major psychoanalytic theorists (Stolorow & Atwood, 1979). It offers a unique means of elucidating the complex and multifaceted reality of both individual lives and the psychoanalytic process.

Clinical case studies of religion seem rare in psychoanalysis. As Pruyser (1973) notes, Freud never published a "full fledged case study that focused on the dynamics of religion in the life of the person" (p. 252). Roland's (1988) discussion of a seriously practicing mystic seen in short-term psychoanalytically oriented treatment is the only case that I have encountered in the literature other than my own earlier work (cf. Rubin, 1992). There is thus a dearth of clinical data regarding religion, especially Asian religion, within psychoanalysis. This chapter attempts to begin to remedy this state of affairs.

In this chapter, I will use the case study method and examine clinical data from a psychoanalytic treatment with a Buddhist meditator. These data, viewed through the lens of psychoanalytic self psychology and object relations theory, will seek to demonstrate that religion is not inherently or monolithically pathological or constructive. Rather, it has multiple meanings and functions involving constructive, self-protective, pathological, reparative, and restitutive components. It may (1) cultivate enhanced self-observational capacities and thus heighten self-awareness; (2) express urgent wishes and aspirations; (3) offer guidelines for morally acceptable behavior, provide a rationale for self-punishment, and reduce self-recriminative tendencies; (4) enhance one's efforts to cope with difficulties or crises; (5) protect, repair, or restore self-representations, that is, enduring images of self, which have been threatened or damaged; (6) facilitate self-demarcation and enhance affect regulation and tolerance; (7) impede awareness of disturbing thoughts, feelings or fantasies; or (8) foster "deautomatization" of thought and action.

Psychoanalytic case material is notoriously complex. In *Narrative Truth and Historical Truth*, Donald Spence (1982) has illuminated the inherent hermeneutical difficulties of the analytic encounter. The apparently "self-evident" data of psychoanalytic treatments are, in actuality, enormously complex and indissolubly connected to the theoretical frame of reference and clinical orientation of both patient and analyst. Reconstruction may often be a contemporary construction: acts of discovery, moments of creation, historical truth, narrative fit. In a compatible vein, Schafer (1983) notes that there is no single, comprehensive psychoanalytic life history of an analysand. Rather, there are various histories of the patient's past shaped by the therapeutic environment, including the intersubjective context of the therapeutic relationship, the personal dynamics of both the analyst and the analysand, and the theoretical orientation and treatment approach employed by the analyst. Thus we need to replace the notion of *the* history of the analysand with the *multiple* histories of the analysand.

I shall confine myself to those details of the case material that illuminate the patient's relationship to religion. I will be employing a strategy of discovery not of proof (Smith, Bruner, & White, 1956). I hope to raise and spark new questions and suggest potentially promising lines of future research.

CLINICAL MATERIAL

Steven, a man in his mid-20s, sought psychoanalytic treatment because of periodic bouts of mild frustration about his career, self-esteem issues, and as part of a more extensive quest for self-development and perfection. When he began analysis he was attending graduate school and pursuing an advanced degree in the social sciences. Involved in many academic projects, his ambition frequently exceeded his output. Although judged competent and successful by peers and students, he had anxiety about his capacities and often felt flawed and inadequate. He usually became involved with women who were accomplished and "difficult." They turned out to be both full of potential and problems. He generally played the role of "caretaker," establishing an environment in which their needfulness was a central focus of the relationship. He often came to resent his role of "healing wounded sparrows."

He had several close male friendships and many acquaintances. His friends seemed bright, kind, and psychologically minded. The friendships involved sharing an interest in psychological, intellectual, and athletic pursuits. These relationships seemed substantial and enduring.

In graduate school Steven became very interested in Asian thought. He read widely in Asian philosophy and psychology and meditated on a regular basis. At least once a year he participated in intensive 2-week-long residential meditation retreats in which he engaged in silent meditation for 10–15 hours a day and followed traditional Buddhist ethical guidelines such as refraining from killing, lying, stealing, sexual misconduct (defined by the school of Buddhism he was affiliated with as not being involved in adultery), and the consumption of mind-altering intoxicants. Outside of the retreat he continued to abide by these principles. He practiced Vipassana or insight meditation, the core technique of Theravadin Buddhism, which I have described earlier.

There are three interconnected dimensions to this practice: the cultivation of (1) ethical purity, (2) mental clarity, and (3) insight and wisdom. Training in ethics is usually the first stage of this practice. Classical Buddhist texts maintain that unethical behavior is motivated by and promotes mental states such as greed, anger, and egotism that disrupt the mind and interfere with developing mental clarity and insight. Following ethical guidelines leads to mental "purification," in which counterproductive behaviors are gradually eroded and mental attentiveness and clarity are cultivated. Mental clarity facilitates the development of insight into the nature of selfhood and wisdom concerning oneself and the world.

Buddhism values nonattachment, voluntary simplicity, selfless behavior, and the cultivation of compassion and equanimity. The view of health underlying traditional Western psychiatry and psychology is, juxtaposed with the Buddhist model, suboptimal. Buddhism claims that there are possibilities for health that exceed the limits of anything discussed in Western psychology.

Initially Steven was enthralled with Buddhist philosophy and psychology. He felt that Buddhism's emphasis on self-awareness, honesty, and generosity was a welcome improvement over the excessive materialism, selfishness, and hypocrisy of daily life. He maintained that daily life, with its emphasis on individualism, promoted endless striving, stress, and emotional discontent. According to Steven, meditation encouraged seeing and appreciating other dimensions of existence, which led to deepened compassion and morality.

He valued psychoanalysis but believed that Buddhist psychology offered a richer view of the workings of mental life than Western psychology. Buddhist models of ideal mental health, e.g., Enlightenment, particularly intrigued him. He maintained that this vision of health transcended the more restricted views of health in psychoanalysis. In his experience, Buddhist teachers embodied higher degrees of insight, morality, and wisdom than the psychoanalysts he had met or read about.

Steven was the oldest of two siblings in a middle-class family. He was 2 years older than his sister. His parents were in their mid-30s when he was born. His parents were atheists and Steven was raised as a secular humanist.

He described his parents as intelligent people who related to each other in an affable but superficial manner. From Steven's account, childhood was a time in which he did well in school, had several close friends, and was well-liked by adults and peers. He described himself as a sensitive child who spent hours playing baseball and touch football with classmates and neighborhood children and devouring biographies of professional baseball and football stars. He seemed to gain inspiration from the lives and training strategies of these sports figures.

As a child, Steven and his mother were quite close. Initially he described her as "kind" and "saintly." He remembered her as a sensitive and curious woman who valued him very much. He had fond memories of the time they spent together in his childhood. In the course of analysis he recovered childhood impressions that she was "controlling" and "infantilizing" and used him as though he was an extension of herself. She "wanted me to do what she wanted me to do.... Everything had to always be her way."

Their relationship changed for the worse when Steven was an adolescent. At that time family life began revolving around the plight of his troubled and enigmatic younger sister. His sister was constantly involved in some sort of trouble, usually involving school or the authorities. Hostile and defiant, she seemed to be deeply invested in rebelling against whatever anyone expected of her. She fought with teachers, took various drugs including marijuana and cocaine, and openly defied Steven's parents.

She was thrown out of several schools, was sent to several psychiatric institutions, subsequently got pregnant, and was in jail several times for drug-related activities. Steven felt that family life was dominated by his sister and that he was neglected.

In the course of analysis Steven's impression of his mother altered. He came to feel that she was an anxious, overly protective woman who was deeply concerned about the opinions of others and avoided conflict at any cost. He felt that she viewed him as if he were an extension of herself and demanded that he conform to her vision of what he should be. He expended a great deal of effort molding himself into the sort of person that would make her feel proud and successful as a parent.

Steven felt that she recruited him to assist her in handling his sister. Steven became the family "moderator," adjudicating between his sister's hostile responses toward her parents and his angry and baffled parents.

For his mother he became a kind of a surrogate husband, providing advice and support about how to handle her daughter and ministering to her emotional needs. As treatment progressed Steven became aware of resenting that he had been cast in the role of family "redeemer." His parents and especially his mother had expected him to become accomplished and provide vicarious glory for her and the family so as to compensate for her sense of herself as a parental failure.

In Steven's view, this expectation led her to view him in an unrealistically grandiose light. When he was in high school, Steven recalled telling his mother that he imagined it would be a wonderful experience to participate in the Olympic marathon. Her response to him—an above-average quarter miler on the high school track team—was, "Why *don't* you run in the next Olympic marathon? *You* can do it."

Steven felt that he had a distant and emotionally depriving relationship with his father, who he described as a competent, critical, emotionally constrained, and perfectionistic small business owner with a bad temper. Steven and his father seemed to share very little. Their conversations revolved around professional sports and Steven's own athletic performance. Steven's father seemed to rebuff his son's attempts to establish a closer bond. After a while, Steven appeared to feel hopeless about communicating with him.

Steven feared that any affectively charged situation might ignite his father's temper. He described his father as susceptible to severe periodic emotional outbursts accompanied by loss of temper, yelling, and screaming. His father tended to become angry or panic when things did not conform to his rigid expectations. In one particularly traumatic incident when Steven was in junior high school, his father spontaneously hit him, without an explanation, after learning that Steven had fought a classmate who had repeatedly insulted his sister. Afterward, Steven made a secret vow to himself that he would never lose his temper and become an "animal" like his father.

His father placed a great premium on performance. Steven presented various memories of his father criticizing and correcting his school work or athletic performance. Steven recounted with sadness that his father's only response after watching him play a nearly flawless game of baseball in junior high school in which he got several hits and made no errors was "you played alright but in the ninth inning you made a mistake." His father was like a "fun house mirror"; whatever Steven did, the reflection was negative (Baker & Baker, 1987, p. 4). Steven felt that he could never please his father and believed that if he was not perfect, then his father viewed him a failure.

Although Steven hated his father's criticalness, he ultimately had a more difficult time deciphering his mother's intensely conflictual communications and actions. He felt that her pseudo-mutuality and interpersonal mystifications substituted for genuine emotional closeness, intrusiveness masqueraded as caring and concern, and hostility was expressed under the guise of "sweetness."

Steven felt that he had subordinated himself to his parents' demands to such an extent that he had sacrificed and lost touch with his own needs and goals. He felt periodically confused about his own goals and ideals and experienced an inordinate amount of self-doubt and self-blame. He had great difficulty assessing his talents and abilities, alternating between his father's belief that he could do nothing and his mother's view that he could do anything. He strove to be supersuccessful so as to compensate for his sister's difficulties and his parent's sense of failure and to win his father's approval. But his childhood and adolescence were pervaded by the sense that he had failed his entire family: he had not cured his troubled sister, redeemed his mother, or pleased his father. He never felt admired for his own accomplishments, which he kept essentially hidden so as not to threaten his sister's precarious self-esteem. He felt he would never live up to his father's ambitions for him, and that his father's emphasis on his improvement reflected a deep disappointment with him.

For the first several years in treatment he experienced his parents as moral exemplars, agreeing with their assessment that they were above moral reproach and essentially justified in criticizing the "mediocre" outside world. He accepted their view that many of his sister's difficulties with the law were due to prejudiced or rigid authorities. Their response after learning that their daughter was arrested when marijuana was found in her car was that "somebody probably planted it."

In the course of treatment Steven came to feel that his parents were hypocritical and self-serving. They seemed committed to not looking at their own responsibility for both his sister's behavior and its effect on Steven. Family life was permeated by half-truths and self-justifications. Given his own commitment to honesty and self-reflection, Steven felt increasingly alienated from his family and beset by self-doubts about the validity of his own goals and ideals.

Although he chose a very different life course from his family and had separated from them in certain ways, the impact of his family was preserved into his adulthood. In his relationships with women, self-abnegation and deference were assumed to be the best way to capture the woman's attention and to sustain her interest. Although he was able to draw on his considerable emotional and intellectual resources to create a meaningful life and achieve professional success, nevertheless his divided loyalties to family members complicated and undermined his self-esteem and his career. He felt intense conflicts between his own upwardly mobile intellectual ambitions and his deep loyalty to his imprisoned sister. He felt torn between the competing demands of not being successful and thus remaining loyal to his sister and his father's demands for perfection. Intellectual success raised the specter of separation, disloyalty, and danger.

In relation to his analyst, Steven was at first somewhat guarded and compliant. Although he seemed to speak with a minimum of self-censorship and inhibition, he was fearful that his analyst would be judgmental, controlling, and intrusive, tacitly or covertly imposing his own agenda on Steven.

Above all Steven wished, as he later put it, "not to rock the boat," not to assert angry or oppositional feelings that would create friction between himself and his analyst. Steven alternated between fear that his anger would lead to his analyst attacking him (like his father) or crumbling (like his mother). As his fantasies about these twin dangers were explored within the transference, his need for an idealized analyst whose insight and wisdom he could utilize and incorporate emerged.

As the analysis proceeded, material about Buddhism periodically emerged. At first Steven treated Buddhism like a historic building in a changing neighborhood and accorded it a kind of "landmark status" in which it was exempt from critical examination and the threat of demolition and preserved in its original form. Gradually, curiosity about Buddhism replaced veneration.

DISCUSSION

We gradually came to understand that Buddhism had taken on constructive, defensive, reparative, and restitutive functions for Steven, simultaneously enriching and limiting his life.

Cultivation of Self-Observational Capacities

Daily life is often pervaded by "mindlessness" or inattentiveness and automaticity of thought and action. Our typical mode of perception is, to a greater extent than we realize, selective, distorted, and outside our voluntary control. A conscious and unconscious blend of anticipatory fantasies, associations, and repetitive concerns often make us unaware of the actual texture of our experience.

In a previous publication (Rubin, 1985) and in the next chapter, I describe the way meditation cultivates increased mindfulness or attentiveness. In summary, I claim that Freud defined the optimal state of mind for analysts to listen to patients as "evenly hovering attention," but he did not discuss, in a positive sense, how to develop it. He focuses on what to avoid (e.g., censorship and prior expectations), not what to do, that is, how to cultivate "evenly hovering attention." Since Buddhist meditation cultivates exactly this state of mind, it can enrich psychoanalytic listening. Steven demonstrated a tremendous ability to access and describe nuances of his thoughts, feelings, and fantasies. Meditation practice seemed to cultivate this unusual degree of self-awareness.

Meditation Reduces Self-Recriminative Tendencies

This increased awareness facilitated greater access to formerly unconscious material. To cite one example among many, on several occasions while meditating, Steven became aware of the formerly unconscious hurt and rage he felt about the way his parents made him feel responsible for family difficulties and the extent to which they encouraged him to fulfill their own needs and goals.

His stance toward these feelings also changed. Nonjudgmental attentiveness—the ability to experience thoughts and feelings impartially—replaced his perfectionistic father's criticalness. As his capacity for empathic self-observation increased, self-recriminative tendencies declined. Thoughts and actions that formerly provided ammunition for him to prosecute himself no longer tended to upset him. As he gradually experienced decreased self-criticalness and self-punishment, he came to feel more patience and compassion.

Self-Demarcation and Affect Regulation

Self-demarcation and affect regulation are two of the building blocks of self-organization and development. Margaret Mahler's (1974) seminal work highlights the crucial importance of the former, particularly the sense of oneself as a demarcated and distinctive person with a unique emotional life and a set of personal ideals and goals. Recent research in infant development in general and the patterning of early infant–caregiver interactions (Stern, 1985) has confirmed that affect regulation is of central importance in the development and organization of self-experience. For Stern (1985), affectivity is a "self-invariant," contributing during the first months of infancy to the development of "the core sense of self" (p. 69). "The rudiments of the infant's sense of self," as Stolorow, Atwood, and Brandchaft (1987) note, "crystallize around its recurrent affect states" (p. 67). Significant caregivers play a central role in encouraging the development and consolidation of self-experience and the interactional context in which the development occurs.

Steven's parents had provided an environment inimical to the development of both of these capacities. Self-demarcation and affect integration had been derailed by their narcissistic use of him. Discovering his own unique goals and needs was difficult in an environment in which he felt compelled to become the sort of person that his parents demanded. As he gradually experienced decreased self-criticalness and self-punishment, he experienced less depressive affect and a clearer sense of his own thoughts, feelings, and goals. This was illustrated by both his emerging sense of his own unique ideals and aspirations and his changing reaction to negative thoughts and affects. He now had a greater capacity for self-soothing. He was able to acknowledge and tolerate painful affect rather than avoid it (like his mother) or panic or become enraged (like his father).

Expression of Urgent Wishes and Ideals/Self-Restoration

Inquiry into the origins and functions of his ambitions revealed that he was haunted by his failure to receive his critical father's approval, save his damaged sister, and redeem his mother. He had a deep fear that to abandon these strivings as compulsive necessities would be to lose forever any possibility of his being important and exciting to his parents (or anyone else).

Because of his father's remoteness and volatility, he never provided Steven with an image of "idealized strength and calmness" (Kohut, 1984, p. 52) that Steven could utilize as a model for cultivating his own goals, ideals, and values. Consequently, this aspect of himself was underdeveloped. He became what Kohut (1971) termed an "ideal hungry personality" seeking to identify with exemplary figures and exalted theories in the external world. In offering an image of and a vehicle for self-perfection, Buddhist meditation afforded him a substitute set of missing ideals and values, thus strengthening a dimension of himself that he needed to fortify to feel good about himself.

These ideals were restitutive and restrictive. Buddhist emphasis on self-purification and transformation had a dual unconscious function: It provided a means of attempting to win his perfectionistic father's approval and atone for his unconscious guilt over his imagined crime of not saving and wishing to destroy his damaged sister. Perhaps if he were perfect, then his father would accept him. Buddhism offered an opportunity to offset his sense of badness and repair the damage he felt he had caused. Buddhism thus became what Melanie Klein (1937), in another context, termed a vehicle for "reparation."

It also became an agent of self-condemnation and self-inhibition. The quest for purity of action, like his father's demand for perfection, became one more ideal that he could never attain, and thus one more occasion for self-condemnation. He periodically spoke of the guilt he felt when he was not meditating on a regular basis or not living up to Buddhism's ethical ideals.

Here, Buddhism led to greater unconsciousness rather than increased mindfulness. Most humans, as Freud aptly noted, gravitate toward pleasure and away from pain. We feel better when we achieve our ideals. We feel badly when we do not attain our ideals. The seeds of self-deception and self-division are planted: our need to feel good will lead us to ignore or bury feedback that we are not acting in accordance with our ideals. A situation is constructed in which nonideal behavior becomes unconscious. Thus, instead of a meditation teacher utilizing the feeling of sexual attraction to a student as feedback about important personal and perhaps interpersonal phenomena, these feelings may be denied or acted out (see Chapter 9).

Rationale for Self-Punishment

Steven was psychologically attracted to Buddhism's emphasis on nonattachment, or not being immersed in or negatively addicted to any phenomena, and its inherent asceticism for several reasons. It gratified what Freud (1926), in another context, termed his unconscious guilt and need for punishment for failing his father, mother, and sister. Second, it served as a rationale to not go beyond or have more than his imprisoned sister. A poignant example of this occurred several years ago when he spent the Christmas weekend doing a self-created "meditation retreat" in his home. For several days he turned off his phone, meditated throughout all his waking hours, and had no human contact. His later association to this event was that he had created a "cell." Fashioning his own "prison" was a way of punishing himself and remaining loyal to his incarcerated sister.

Asceticism—having little—also protected him from reliving the trauma of being "robbed" by his sister and unprotected and undefended by his parents. If he had scant possessions, only some "spare change," then less could be taken from him than if he were wealthy and had a "large coin collection."

Deautomatization of Thought and Action and Enhancement of Capacity to Cope with Difficulties or Crises

In his studies of mystical practices and psychotherapy, Arthur Deikman (1982) emphasizes the centrality of "deautomatization," or an undoing of automatized, habitual thought and action. In cultivating perceptual acuity and attentiveness, meditation fostered awareness of and deautomatization from previously habitual patterns. This led Steven to feel an increased freedom of action while also lessening the potential for the acting out of unconscious negative affects associated with danger to himself and his family. Being deautomatized minimized the chances of his being out of control like his punitive father. Thus, there was less chance of him acting like an "animal" or anyone getting "whipped."

Religion, according to Ludwig Feuerbach, the nineteenth-century German philosopher, is a form of human self-alienation. It gradually became evident that Buddhism's emphasis on cultivating "cool" rather than "hot" emotions, equanimity rather than passion (Kramer & Alstad, 1993), actually inhibited him in certain ways. For example, it reinforced his defensive passivity. In attempting to develop such qualities as equanimity and compassion, Steven focused on detaching from negative affects rather than experiencing them. This blocked the emergence of moral outrage against his parents for neglecting his needs and allowing his disturbed sister to dominate family life. The possibility of Steven being appropriately assertive or angry was thus unfortunately stifled.

Various unconscious meanings and purposes emerged and his image of and relationship to Buddhism altered in the course of treatment. From our perspective he developed a more nuanced and critical perspective on Buddhism. He found its vision of reality, a world everywhere beset by suffering caused by internal psychological attachment, to be an incomplete diagnosis of human suffering and worthy of critical scrutiny. Now he viewed Buddhism not as absolute ahistorical, transcultural Truth but as what Schafer (1976) has termed, in a different context, a "vision of reality" (p. 23).

Buddhism's vision of reality seemed, to him, to neglect intimate relationships, the body, the sociocultural causes of human suffering, the influence of the past on current behavior, and the way transference and resistance shape and delimit ordinary awareness and the quality of one's life.

He also felt that Buddhist goals such as complete liberation were unrealistic and psychologically enslaving. He believed that ideals such as complete awareness and total eradication of self-centeredness result in self-division and self-contempt.

Buddhism's unattainable ideals became, for Steven, further grist for the self-condemnation mill. He realized that seeking Enlightenment was like playing baseball in front of his father: He could not win.

Roland Dalbiez (1936) distinguishes between a system's doctrine or its theories, and its method, or its operationalizable techniques. In the course of treatment, Steven's interest in Asian psychology shifted from its theory to its method—the practice of meditation, the training of attentiveness, and the mind observing its own workings—rather than Buddhist viewpoints on such things as rebirth, the etiology of suffering, or the possibility of completely transcending psychological conditioning.

He also began to deidealize Buddhist teachers. Admiration rather than idealization and subordination now characterized his relationship with them. He no longer viewed them as sources of complete insight and wisdom but as mortals whose commitment to self-examination and ethicality he deeply respected.

From my perspective these changes were an outgrowth of both analytic and meditative experiences. As the idealizing self–object transference developed toward me, Steven's idealizing tie to Buddhism decreased. My ongoing commitment and availability and my patience in not prematurely challenging Buddhism's "landmark status" seemed to be important in order not to reenact Steven's relationship with his controlling mother, who utilized him to serve a self–object function for *her.* Understanding that exploration was experienced at this point in the treatment as an intrusion helped me temper my countertransference-based desire to prematurely illuminate the unconscious meanings of Buddhism. This stance ensured that Steven's fear that I would be like his mother and control him by telling him what to think or how to be was thus not actualized.

Idealization was replaced by a more curious and playful stance toward Buddhism. As the analysis proceeded, we treated Buddhism more like a dream than a sacred monument, examining his associations to it rather than assigning it a standardized meaning. We attempted to illuminate its unconscious meanings and purposes rather than assume that it was either inherently pathological or unworthy of psychoanalytic scrutiny.

As the therapeutic tie strengthened, a new self–object transference, what Stolorow and Atwood (1990) have recently termed the "self-delineating" self–object transference emerged. What seemed of greatest motivational priority at this juncture of the transference was the delineation and acceptance of Steven's thoughts and feelings. Gradually Steven began acknowledging a greater range of negative affects such as anger, disappointment, and rage. He began associating to various unconscious aspects of his participation in Buddhism and connecting them to dynamic issues arising in treatment. For example, he connected his tendency to minimize or disavow feelings of disappointment or emotional deprivation with his interest in Buddhism's advocacy of detachment and asceticism. We saw that Buddhism sometimes became a rationale to attempt to be kind and nonjudgmental to people who were mistreating or exploiting him. This stance impeded the emergence of feelings of disappointment or anger.

As these and other related feelings emerged, Steven gradually began questioning other dimensions of Buddhist theory such as its account of self-experience and the etiology of suffering. The increasing trust in his own experience facilitated by the self-delineating self–object transference facilitated a more careful examination of his own experience with and critique of Buddhist doctrine. As I indicated earlier, he gradually developed a more nuanced perspective on Buddhism in general and his own participation in it in particular. This was demonstrated by his challenging of experience-distant aspects of the theory such as rebirth and his utilization of heuristic aspects of the method and the theory. He now seemed to sail a course between the Scylla of unbridled reverence and the Charybdis of defensive renunciation.

CONCLUSION

... dreaming is the key to the mysteries of religion.
—Ludwig Feuerbach

From its inception, psychoanalysis has aspired to the status of science. In *New Introductory Lectures on Psycho-Analysis*, Freud (1933) makes this explicit, asserting that science is psychoanalysis' Weltanschauung, or worldview. Scientists have waged a long battle to obtain freedom from religious control. It is thus natural that religious matters would be suspect within psychoanalysis.

Numerous incidents in the history of religion past and present, ranging from the Inquisition to recent scandals involving fundamentalist preachers and Buddhist meditation teachers, lend support and give us no reason to question Freud's and psychoanalysis' interpretations of the pathological nature of religion. We may thus be tempted to dismiss religious phenomena as psychologically immature or irrational.

Challenges to psychoanalytic reductionism have often fallen victim to a reverse pitfall: accepting religious experience too uncritically. If Freud, as well as many subsequent psychoanalysts, precipitously dismissed religion, then much theological discourse and some nonpsychoanalytic discourse that is sympathetic to religion's capacity to enrich human life tend to uncritically accept it. The complex conscious and unconscious processes by which people come to possess and use religious imagery are ignored in both approaches to religion (J. McDargh, 1989, personal communication).

The problem is that not all religious involvements are inherently pathological or unequivocally constructive. Religion, like all psychic events, has multiple meanings and functions (Waelder, 1936).

Freud's revolutionary perspective on dream interpretation provides an indispensable starting point in illuminating these meanings and functions. The meaning of a dream, according to Freud (1900), is arrived at not by translating dream material into the a priori meaning and "fixed-key" of a "dream-book" but by eliciting the dreamer's unique associations. The meaning of religious phenomena, it seems to me, is arrived at not by translating it, as so many analysts and theologians have done, into the fixed-key of the analyst's, or theologian's, dream book on religion but by elucidating a specific practitioner's unique associations.

What difference does it make to adopt this approach to religious material? First, one would approach the material asking a different set of questions. Rather than asking what are the pathological meanings and determinants of religion, the clinician would think about its various unconscious meanings and purposes. The analyst might, for example, ask: (1) Does religion facilitate or interfere with increased self-awareness? (2) What wishes or aspirations does it express? (3) What guidelines for acceptable behavior and what rationale for self-punishment are presented? Does religious involvement reduce or increase self-recriminative tendencies? (4) Does religion enhance one's ability to cope with difficulties or crises? (5) What effect does religion have on the patient's self-experience? Does it protect, restore, or repair self or object representations? (6) Does it impede awareness of disturbing thoughts, feelings, or fantasies? This is by no means a complete list of the questions that might be asked about religious phenomena. Other clinicians will undoubtedly discover other salient meanings and functions.

Such a strategy would help psychoanalysts and religionists avoid the twin dangers of unbridled reverence, which accepts religion without critical examination, or a priori pathologizing, which rejects it automatically. The picture of Buddhism that emerges from such an approach is that of a complex mosaic involving constructive, pathological, restitutive, integrative, and transformative dimensions.

It is fallacious to equate religious involvement with psychological immaturity or religious disbelief with psychological maturity. Each person's ultimate decision to participate in religion and their relation to it is unique and can be revealed only through detailed individual study. Religious involvement can best be understood not by either a priori pathological interpretations or uncritical acceptance, but rather by examining its unconscious meanings and functions in a practitioner's life. The task of the clinician is to ascertain the unconscious meanings, functions, and motivational priority for a particular patient at a particular time.

REFERENCES

Baker, H., & Baker, M. (1987). Heinz Kohut's self psychology: An overview. *The American Journal of Psychiatry*, *144*(1), 1–9.

Dalbiez, R. (1936). *La methode psychanalytique et la doctrine Freudienne* (2 Vols.). Paris: Desclee de Bouwer.

Deikman, A. (1976). *The observing self: Mysticism and psychotherapy*. Boston: Beacon Press.

Freud, S. (1900). The interpretation of dreams. In J. Strachey (Ed. & Trans.), *The standard edition of the complete psychological works of Sigmund Freud*, Vols. 4 & 5. London: Hogarth Press.

Freud, S. (1926). Inhibitions, symptoms and anxiety. In J. Strachey (Ed. & Trans.), *The standard edition of the complete psychological works of Sigmund Freud*, Vol. 20 (pp. 77–175). London: Hogarth Press.

Freud, S. (1933). New introductory lectures. In J. Strachey (Ed. & Trans.), *The standard edition of the complete psychological works of Sigmund Freud*, Vol. 22 (pp. 5–182). London: Hogarth Press.

Klein, M. (1975). *Love, guilt & reparation*. London: Hogarth Press.

Kohut, H. (1984). *How does analysis cure?* Chicago: University of Chicago Press.

Kohut, H. (1971). *The analysis of the self*. New York: International Universities Press.

Kramer, J., & Alstad, D. *The guru papers: Marks of authoritarian power*. Berkeley, CA: North Atlantic Press.

Mahler, M. (1974). *The selected papers of Margaret Mahler, Vol. 2*. New York: Jason Aronson.

Nyanaponika, T. (1972). *The power of mindfulness*. San Francisco: Unity Press.

Pruyser, P. (1973). Sigmund Freud and his legacy: Psychoanalytic psychology of religion.

In C.Y. Glock & P.E. Hammond (Eds.), *Beyond the classics* (pp. 243–290). New York: Harper & Row.

Roland, A. (1988). *In search of self in India and Japan: Toward a cross-cultural psychology.* Princeton, NJ: Princeton University Press.

Rubin, J.B. (1985). Meditation and psychoanalytic listening. *Psychoanalytic Review, 72*(4), 599–613.

Rubin, J.B. (1992). Psychoanalytic treatment with a Buddhist meditator. In M. Finn & J. Gartner (Eds.), *Object relations theory and religion: Clinical applications* (pp. 87–107). Westport: Praeger.

Schafer, R. (1983). *The analytic attitude.* New York: Basic Books.

Schafer, R. (1976). *A new language for psychoanalysis.* New Haven, CT: Yale University Press.

Smith, M., Bruner, J., & White, R. (1956). *Opinions and personality.* New York: John Wiley.

Spence, D. (1982). *Narrative truth and historical truth.* New York: Norton.

Stern, D. (1985). *The interpersonal world of the infant.* New York: Basic Books.

Stolorow, R.D., & Atwood, G.E. (1979). *Faces in a cloud: Subjectivity in personality theory.* New York: Jason Aronson.

Stolorow, R.D., & Atwood, G.E. (1990). Lecture on October 20, 1990, at the Thirteenth Annual Conference on the Psychology of the Self, New York, New York.

Stolorow, R.D., Atwood, G.E., & Brandchaft, B. (1987). *Psychoanalytic treatment: An intersubjective approach.* Hillsdale, NJ: Analytic Press.

Waelder, R. (1936). The principle of multiple function. *Psychoanalytic Quarterly, 35,* 45–62.

III

The Buddhist and Psychoanalytic Process

6

Meditation and Psychoanalytic Listening

One is bound to employ the currency that is in use in the country one is exploring.

—Freud (1911)

In scientific affairs there should be no place for recoiling from novelty.

—Freud (1924)

The history of science is rich in the example of the fruitfulness of bringing two sets of techniques, two sets of ideas, developed in separate contexts for the pursuit of truth, in touch with each other.

—Robert Oppenheimer (1954)

It was Freud (1900) who taught us that listening to ourselves and our patients is both the essential tool of psychoanalytic inquiry and the foundation of psychoanalytic technique.[1] Listening is crucial to all that the analyst does. Every aspect of the therapeutic process—hearing, associating, imagining, empathizing, hypothesizing, formulating, interpreting, intervening, and validating—begins with and is predicated on listening.

Freud's (1912) recommendations for cultivating "evenly hovering attention" (pp. 111–112) and thus facilitating optimum listening have been unanimously accepted as the cornerstone[2] of the listening process

---

[1]For reasons of expediency, I shall use the words "psychoanalyst" and "psychoanalytic" in a generic sense to include all forms of therapists and therapies that rely on dialogue rather than medication, behavior modification, and the like.

[2]I am not going to discuss the relationship between evenly hovering attention and the more selective and focused mode of perception that is required for other aspects of the analyst's and analysand's functioning. These modes of perception are essential to the psychoanalytic process, but evenly hovering attention is the *sine qua non* of all aspects of listening. In order to formulate, interpret, or validate, one first needs to listen with evenly hovering attention.

in the standard texts on psychoanalytic technique (Fenichel, 1941; Glover, 1955; Greenson, 1967; Sharpe, 1930) and the subsequent psychoanalytic examinations of listening and the analyst's state of mind (Balter, Lothane, & Spencer, 1980; Bion, 1967, 1970; Brown, 1977; Chrzanowski, 1980; Cohen, 1952; Ferenczi, 1928, 1930; Fliess, 1942; Gray, 1973; Grotjahn, 1950; Heimann, 1977; Khan, 1977; Langs, 1978; Little, 1951; McLaughlin, 1975; Olinick, 1980; Reich, 1966; Searles, 1959; Sterba, 1934). And yet, despite the indispensability of "evenly hovering attention" in the listening process, the most crucial pragmatic question associated with it—how to cultivate it—has been neglected and incompletely understood by Freud and subsequent psychoanalytic writers. Freud (1900, 1912) recommended what psychoanalysts need to do—listen with "evenly hovering attention"—but he did not reveal, in a positive sense, how to do this. He presented a "negative" account—what to avoid (e.g., censorship, prior expectations, and "reflection")—not what to do, that is, how to cultivate "observation" and "evenly hovering attention." Freud left unanswered two questions: (1) Can the capacity to listen with "evenly hovering attention" be cultivated systematically? (2) If so, how can this be done?

Freud's perspective has been uncritically accepted in almost every subsequent psychoanalytic discussion of listening. Bion's (1970) remarks on forsaking "memory, desire and understanding" have been the only attempt in the literature to supplement Freud's formulations. No one has actually refined or enriched Freud's remarks. The psychoanalytic community has not, for the most part, either recognized the possibility of developing this refined capacity to listen or directly focused on its cultivation. Authors who discuss "evenly hovering attention" either simply emphasize the importance of listening in this special way (Altman, 1969; Balter et al., 1980; Bion, 1970; Greenson, 1967; Langs, 1978; Levenson, 1972; Stolorow & Lachmann, 1980) or merely imply that it can be developed and enhanced (Balter et al., 1980; Bion, 1970; Stolorow & Lachmann, 1980). No systematic procedure and no specific recommendations have even been offered to cultivate "evenly hovering attention." This omission has hampered the optimum development of psychoanalytic listening.

Attempts to cultivate optimal listening were not the sole province of and did not originate with Freud or Western psychology. Analytic examinations of listening are only one version, and in fact a comparatively recent and incomplete edition, of an enterprise that has been pursued for thousands of years in various Eastern psychological systems and preserved with lapidary precision in the Theravadin Buddhist meditative tradition (Goldstein, 1976; Goleman, 1976; Kornfield, 1977; Mahasi, 1973, 1978; Thera, 1941), a nontheistic mental and attentional training system. In this chapter, I shall examine the meditative technique taught in the Theravadin tradition, which serves as a prototype of the 12 extant meditative systems (Goleman, 1977).

Meditation has been virtually ignored and devalued in psychoanalysis. Very little has been written about it in the psychoanalytic literature.[3] Extensive documentation about meditation exists; there are comprehensive phenomenological and theoretical descriptions (Buddhaghosa, 1976; Goldstein, 1976; Goleman, 1977; Kornfield, 1977; Mahasi, 1973, 1978; Nyanaponika, 1962, 1972; Suzuki, 1970; Thera, 1941, 1962, 1972; Walsh, 1980; Welwood, 1979; Wilber, 1980), as well as recent research data validating that meditation can alter and extend apparently involuntary processes such as attention and listening beyond previously established limits (Kornfield, 1979). Unfortunately, the available psychoanalytic discussions (e.g., Alexander, 1931; Carrington, 1981; Carrington & Ephron, 1975; Shafii, 1973) have not fully attended to or recognized the transformative possibilities of classical meditation.

Understanding meditation, like comprehending psychoanalysis, requires personal experience and theoretical knowledge. Several Western behavioral scientists who were originally skeptical about meditation have indicated that positive claims about meditation, which made little sense prior to personal exposure, only became intelligible after meditating (cf. Walsh, 1980). Since most analysts are not familiar with either the meditative literature or practice, discussions of meditation are permeated by corrosive emotional biases and intellectual preconceptions. Accurate assessment demands that meditation be disengaged from the plethora of distortions and erroneous associations that surround it. Meditation is not religious dogma, hypnosis, mysticism (Masson & Masson, 1978), regression, or pathology (Alexander, 1931).

Masud Khan's (1977) description of a state of mind he terms "lying fallow" (p. 397) begins to paint a verbal picture of the meditative state of mind for the nonmeditator. Meditation, like "lying fallow," is "alerted quietude and receptive wakeful lambent consciousness" (p. 397). Meditation is a generic term for a psychologically cogent and incisive corpus of theories and techniques of mental training, originating 2500 years ago in India and later developed in China, Japan, and various Far Eastern countries. Devoted to the cultivation of nonexclusionary perception and the ability to observe and listen with nonselective and nonrestrictive awareness, meditation can be of immense value to psychoanalysis in

---

[3]"Meditational" techniques have been utilized in various nonanalytic therapies, such as Gestalt, psychosynthesis, and autogenic training.

general[4] and psychoanalytic listening in particular, providing a systematic and efficacious technique for cultivating precisely the capacity and state of mind that Freud recommended for optimum listening. Whereas Freud suggested what we need to do, the meditative tradition reveals how to do it. Thus, meditation can be utilized to enrich the psychoanalyst's capacity to listen.

In order to suggest how meditation can do this, I shall (1) examine the salient aspects of Freud's theoretical recommendations for facilitating optimum listening and Bion's attempted addendum, particularly his recommendations for eschewing "memory, desire, and understanding"; and (2) explore the meditative tradition's method for developing "evenly hovering attention."

FREUD

Freud's recommendations for facilitating optimum listening, which served as the foundation for the subsequent psychoanalytic view of listening, were summarized in several incisive passages in *The Interpretation of Dreams* (1900) and "Recommendations to Physicians Practicing Psycho-Analysis" (1912). The theoretical linchpin of his recommendations for facilitating optimum listening is in his instructions to analysands on how to prepare for dream interpretation. Two changes are required: (1) "an increase in the attention he pays to his own psychical perceptions," and (2) "the elimination of the criticism by which he normally shifts the thoughts that occur to him" (1900, p. 101). For attention to increase and criticism to decrease, there must be the creation of what I would term the optimum "internal" and "external"[5]

[4]Meditation can reduce self-recriminative and addictive tendencies (Marzetta, Benson, & Wallace, 1972); lower sensory thresholds (Walsh, 1980); increase imperviousness to distractions; decrease automatization and selectivity of perception; enhance perceptual sensitivity and clarity; improve attention deployment; facilitate free association; reduce countertransferential blindspots; and enhance the capacity for empathic self-observation and self-analysis.

[5]"Internal" refers to the listener's state of mind while listening; is one reflecting and thus listening with a spirit of *parti pris* (Freud, 1900, p. 523), or observing (Freud, 1900, p. 101) and listening with "evenly hovering attention"? I shall discuss this crucial distinction in the next section. "External" refers to the analytic environment, especially the quiet, tranquil atmosphere, which is arranged to minimize and decrease visual, auditory, and kinesthetic stimulation and diversions. To create the optimum external environment, Freud (1900, 1904) recommended limiting and reducing physical movement—the analysand lies in a restful position on a couch—and minimizing access to and thus reducing external stimuli and distractions during the session. Decreasing distracting phenomena creates an atmosphere of greater tranquility that facilitates increased attentiveness to internal processes (Freud, 1904).

environment. To understand how the internal environment is created, it is essential to understand Freud's (1900) distinction between two ostensibly similar states of mind, "reflecting" and "observing."

‡Reflecting is a selective and discriminatory mode of perception, a state of mind in which one edits, rejects, and suppresses certain thoughts and feelings instead of being nonjudgmentally aware of them. This perceptual filtering distorts observation and impairs listening. Observation, on the other hand, is an impartial state of mind in which one notes what is without criticism or censorship‡ Freud's description of this state of mind is worth repeating:

> It ... consists in making no effort to concentrate the attention on anything in particular, and maintaining in regard to all that one hears the same measure of calm, quiet attentiveness—of "evenly-hovering attention" ... as soon as attention is deliberately concentrated in a certain degree, one begins to select from the material before one; one point will be fixed in the mind with particular clearness and some other consequently disregarded, and in this selection one's expectations or one's inclinations will be followed. This is just what must not be done ... if one's expectation's are followed in this selection there is the danger of never finding anything but what is already known, and if one follows one's inclinations anything which is to be perceived will most certainly be falsified. (1912, pp. 111–112)

The actual listening process involves creating a tranquil, analytic environment and then listening with "evenly hovering attention."

⎨If Freud's recommendations are followed, then receptivity to mental phenomena is enhanced and attentiveness and listening become refined. One notices thoughts, feelings, and fantasies that one ordinarily would not be aware of and one is able to attend to these phenomena more clearly and completely.⎬

Freud had contradictory views concerning the possibility of listening in this manner. In the same text, Freud (1900) asserted both that this attitude is not difficult to achieve—"the adoption of an attitude of uncritical self-observation is by no means difficult" (p. 103)—and that effort and "practice" are needed for it—"practice is needed even for perceiving endoptic phenomena or other sensations from which our attention is normally withheld" (pp. 522–523). While he recognized that one cannot automatically and effortlessly observe oneself and listen with "evenly hovering attention," he also assumed that all one had to do was do it.

Freud's contribution to the topic of listening was threefold: he discovered and described (1) the ideal state of mind for optimal listening; (2) the primary barriers to optimal listening; and (3) the ideal external environment for optimal listening.

Unfortunately, he blurred the distinction between the will to listen with "evenly hovering attention" and the capacity to do so. The difficulty in sustaining attentiveness (James, 1962) and the ubiquity of inattention and compromised[6] listening (Little, 1951; Searles, 1959) calls into question Freud's view and suggests that optimum requires more than merely the wish to listen because it is an arduous task even for specially trained professionals. It requires, and can be refined by, special training and practice. Freud's suggestions for removing the hindrances to listening are an essential element in refining listening, but they do not constitute the whole story. Just as stating what is immoral does not, in and of itself, generate moral action, revealing the obstacles to optimum listening does not, in and of itself, foster listening.

BION

In "Notes on Memory and Desire" and "Opacity of Memory and Desire," Bion (1967, 1970) presents a critique and provides a reformulation of the traditional epistemology of listening. He challenges the consensually held view that desire, memory, and understanding[7] facilitate the listening process and suggests that they serve as filters that interfere with accurate observation and listening. They are "obstacles intervening between the psychoanalyst and the emotional experience of the session" (1967, p. 280), and thus preclude optimum observation and listening. Because their presence is more deleterious than their absence, Bion proposes the shunning of memory, desire, and understanding (1970):

> It is important that the analyst should avoid mental activity, memory and desire which is ... harmful (p. 42) ... the capacity to forget, the ability to eschew desire and understanding, must be regarded as an essential discipline for the psychoanalyst. Failure to practice this discipline will lead to a steady deterioration in the powers of observation whose maintenance is essential. The vigilant submission to such discipline will by degrees strengthen the analyst's mental powers just in proportion as lapses in this discipline will debilitate them. (pp. 51–52)

To achieve this state of mind, Bion (1970) recommends that the analyst should divest the mind of memory, desire, and understanding through

[6]The reality of incomplete and imperfect analysis (Freud, 1937; Schafer, 1976) and the concomitant ubiquity of countertransference, which is, at least in part, selective and compromised listening, serves as a testament to the inevitability of inattentive listening.
[7]"Desire" refers to the wish for understanding, the end of the session, or therapeutic results. "Memory" refers to the remembrance of previous sessions—who the patient *was*. "Understanding" refers to knowledge of psychoanalytic theories and images of who the patient is.

"suppression" (p. 47), "suspension" (p. 46), and "disciplined denial" (p. 41).

While Bion aptly realized that memory, desire, and understanding can interfere with accurate listening, his proposal is flawed, because attempting to control the mind—the effort to eliminate or suppress any mental phenomena—leads to a "reflective" state of mind that precludes "evenly hovering attention" (Freud, 1900). Trying to empty the mind of anything will not result in the mind being empty and receptive. As Seng-tsan, a Japanese teacher of meditation suggested: "When you try to stop [mental] activity to achieve passivity [receptivity], your very effort fills you with [mental] activity" (1984, p. 155). The desire to have no desires (or memory or understanding) is another desire and does not "empty" the mind but keeps it full of and occupied with the thought of (the importance of) being without desires. Trying to overcome a state of mind by artificially altering perception, like trying to get out of a ditch by spinning the wheels of a car, results not in egress but in further entanglement.

"Evenly hovering attention" cannot be attained by the kind of mentally coercive approach that Bion proposes. Rather, one must have perspective openness: nonselective acceptance and awareness of whatever is, right here, now, moment to moment. Dennis Brown's (1977) handling of his drowsiness during a session serves as a heuristic illustration of what is problematic with Bion's method:

> ... ultimately an emerging trust in the understandability of what was happening reduced my anxiety and let me go along with it. Thus I stopped interfering with it by fighting it ... priority was given to Freud's recommendation to aim at "evenly-hovering-attention" over his injunction not to make a determined effort to register and recall [or in Bion's case to forget]. At times the aim became to remain capable of attention by attending to what seemed to prevent it. (pp. 481–482)

MEDITATION AND PSYCHOANALYTIC LISTENING

Although meditation is easy to describe and sounds simple to do, it actually involves states of heightened attentiveness and perceptual acuity with which nonmeditators are ordinarily unfamiliar. A summary of the basic aspects of the meditative process will help the reader understand the ensuing discussion.

In the first stage of the meditative process one sits in a comfortable position in a quiet place, closes one's eyes, and attends to one object, such as the physical sensation of the movement of the stomach. Attention is not thinking about or analyzing what is occurring. It is a simple registering of what is happening. One invariably loses awareness of the abdomen and "wanders" off and "follows" thoughts, feelings, fantasies, and associations. When this happens, one simply notices what is happening without further elaboration or criticism. As soon as one is aware that one is wandering, one returns one's attention to the physical movement of the abdomen. When wandering subsequently occurs, the same procedure is followed. When distractedness decreases and attentiveness becomes more sustained and automatic, stage 2 begins. In stage 2, one becomes attentive in a nonselective and nonrestrictive way to whatever occurs. Instead of maintaining an exclusively focused awareness— being aware of a specific object such as the abdomen—one has an inclusive attentiveness, fully allowing and remaining aware of whatever occurs. Stages 1 and 2 are not mutually exclusive and do not function in precisely the linear way that my description suggests. Stage 1 provides the foundation for the cultivation of "evenly hovering attention" in stage 2 by creating the preconditions for listening with "evenly hovering attention": decreasing distractions, quieting and focusing the mind, and enhancing the capacity to perceive with nonselective attentiveness. Stage 2 utilizes this foundation to cultivate and refine sustained "evenly hovering attention." Without this foundation, it is much more difficult to listen in this specialized manner.

The predominant aspect of stage 1 of the meditative process is the struggle between attentiveness and mental distractedness. Although we believe that we are "aware," and we often respond efficiently and competently to the enormous complexities and demands of daily life, closer inspection reveals that our typical mode of perception is, to an unrecognized extent, selective, distorted, and outside voluntary control (Nyanaponika, 1962). We often operate on automatic pilot, reacting to a conscious and unconscious blend of fallacious associations, anticipatory fantasies, and habitual fears, that make us unaware of the actual texture of our experience. Rarely do we achieve the mental quietness and attentiveness that is a prerequisite for accurate and penetrating listening. We are so habituated to this state of affairs that we are unaware it is occurring. We do not know anything else is possible and we erroneously assume it is the normal state of affairs and the optimum way of listening. A mind that is not specifically trained has difficulty appreciating (1) the extent to which we are unaware; (2) the way this mode of perception distorts and constricts observation and listening; and (3) the possibilities of being qualitatively more aware.

Stage 1 of the meditative process is designed to improve our perceptual acuity and increase our attentiveness. The meditational situation— the decreased mobility and sensory stimulation and the increased concern with internal processes—like the psychoanalytic situation provides the external preconditions for optimum observation. If one continues to follow the basic procedure of stage 1, simply noting the bare facts of perception without any extraneous elaboration, there eventually develops what the meditative tradition terms a "one-pointedness" (Kornfield, 1977) of attention. One-pointedness is not synonymous with what Freud termed "reflection" because it is not dispersed or exclusive perception, but what William James (1962) in another context termed "concentrated attention." One-pointedness of attention is a unification of perception that resembles the focusing of a laser. This state of mind has refined penetrative capacities (Kornfield, 1981) that put a brake on and cut through our skewed mode of perception (Wilber, 1980) and reduce the factors that generate inattention and wandering. This engenders an unaccustomed tranquility that leads to more refined and incisive listening.

Stage 1 of the meditative process creates the essential preconditions for listening with "evenly hovering attention," facilitating the optimum internal and external conditions for listening. The importance of stage 1 is that it offers a heuristic alternative model to its analytic counterpart. The meditative tradition has a greater appreciation of the difficulties in being attentive, and thus places greater emphasis on and provides techniques for training this capacity. More specifically, the meditative model provides a corrective to three crucial issues that Freud and subsequent clinicians ignored:

1. It recognizes that this aspect of the listening process—setting the stage for listening by creating the essential preconditions for listening—was not merely supplemental but indispensable. Although Freud recognized the importance of stage 1, he did not emphasize the complexity and indispensability of this aspect of listening, especially one-pointedness. Without quieting the mind and focusing attention, it is nearly impossible to truly listen.
2. Stage one has an internal as well as an external component. Freud focused on the external environmental conditions that would aid listening—reducing physical mobility and sensory stimulation—but he did not offer recommendations for creating the optimum internal environment (except in the negative sense of eliminating hindrances such as "reflection"). Minimizing hindrances is a necessary but insufficient condition for facilitating optimum listening; optimum listening also requires developing mental tranquility and attentional precision.
3. The optimum internal environment, particularly mental tranquility and attentional precision, can be systematically developed and refined by means of meditation.

Stage 2 of the meditative process is devoted to doing this. In stage 2, one utilizes the awareness developed in stage 1 to cultivate and refine systematically sustained "evenly hovering attention." During this stage, wandering has decreased and attentiveness has increased. One often notices the "first instant" (Tulku, 1975) of perception without getting lost in or carried away by subsequent associative elaborations. One notices phenomena, including one's inattention, more quickly and clearly. This leads to decreased mental "static," increased equanimity, and incisive observational capacities, a state of mind that the Dalai Lama (Gyatso, 1982), in another context, termed "calm abiding." The decreased agitation and increased equanimity lead to more refined and sustained attentiveness. At a certain point, when the attentiveness is sufficiently refined, nonselective awareness ("evenly hovering attention") is more automatic and heightened.

Stage 2 differs from the analytic model in its (1) recognition of the possibility of cultivating refined attentiveness, and its (2) presentation of techniques for doing this. Ordinary attentiveness is only a distant relative of the heightened attentiveness of experienced meditators.

Attempting to listen, investigate, and understand ourselves and others without developing heightened attentiveness is like taking a photograph with a strong and sensitive lens and an unsteady hand: even if the lens (the investigative tool) is refined, if the camera is unsteady (if attentiveness is lacking) the picture (the view of the mind) will be blurred. A grave flaw, and arguably the greatest limitation of the analytic view of listening, has been the lack of understanding of the indispensability of one-pointedness and observation. In the analytic view, one reaches more refined levels of attentiveness by removing barriers. Although this approach has traditionally been accepted, if one attempts to observe and listen without quieting the mind, then one has a greater difficulty becoming aware of phenomena that are accessible when attentiveness is developed and then combined and integrated with it.

There are two ways that meditation can be utilized to enrich analysts' capacity to listen: (1) meditation can be employed in training analysts, and (2) analysts can meditate before or between sessions or at home.[8]

The positive significance of meditation has yet to be appreciated in psychoanalysis. The meditative tradition provides a fourfold contribution to analytic listening: (1) it recognizes that attentiveness can be

[8]I am not going to discuss how meditation could be utilized by patients. Consult Deatherage (1975) and Wortz (1982) for interesting and useful discussions of this issue. Chapters 5 and 9 explore this topic from various angles.

trained, that it is possible to cultivate and refine this capacity; (2) it alters and extends our vision of what is possible in refining the capacity to listen; (3) it provides a specific technique for facilitating optimum listening that both amplifies Freud's seminal theoretical prescriptions for cultivating optimum listening and avoids the limitations in Freud's "negative" model and Bion's coercive approach; and (4) by providing a viable means of actualizing Freud's theoretical recommendations for cultivating optimum listening, it complements and supplements psychoanalytic technique.[9]

REFERENCES

Alexander, F. (1931). Buddhistic training as an artificial catatonia: The biological meaning of psychological occurrences. *Psychoanalytic Review, 18*, 129–145.

Altman, L. (1969). *The dream in psychoanalysis.* New York: International Universities Press.

Balter, L., Lothane, Z., & Spencer, J.H. (1980). On the analyzing instrument. *Psychoanalytic Quarterly, 49*, 474–504.

Bion, W. (1967). Notes on memory and desire. *Psychoanalytic Forum, 2*, 271–280.

Bion, W. (1970). *Attention and interpretation.* New York: Basic Books.

Brown, D. (1977). Drowsiness in the countertransference. *International Review of Psycho-Analysis, 4*, 481–492.

Buddhaghosa, B. (1976). *The path of purification* (B. Nyanamoli, Trans.). Berkeley, CA: Shambhala.

Carrington, P. (1981). Can a spiritual discipline foster mental health? In R. Stewart (Ed.), *East meets West: The transpersonal approach* (pp. 38–55). Wheaton, IL: The Theosophical Publishing House.

Carrington, P., & Ephron, H. (1975). Meditation and psychoanalysis. *Journal of the American Academy of Psychoanalysis, 3*, 43–57.

Chrzanowski, G. (1980). Reciprocal aspects of psychoanalytic listening. *Contemporary Psychoanalysis, 16*, 145–162.

Cohen, M.B. (1952). Countertransference and anxiety. *Psychiatry, 15*, 231–243.

Deatherage, G. (1975). The clinical use of "mindfulness" meditation techniques in short-term psychotherapy. *Journal of Transpersonal Psychology, 7*, 133–143.

Fenichel, O. (1941). *Problems of psychoanalytic technique.* New York: The Psychoanalytic Quarterly.

Ferenczi, S. (1928). The elasticity of psychoanalytic technique. In S. Ferenczi (Ed.), *Final contributions to the problems and methods of psychoanalysis* (pp. 87–101). New York: Basic Books.

Ferenczi, S. (1930). The problems of relaxation and neocatharsis. In S. Ferenczi (Ed.), *Final contributions to the problems and methods of psychoanalysis* (pp. 108–125). New York: Basic Books.

Fliess, R. (1942). The metapsychology of the analyst. *Psychoanalytic Quarterly, 11*, 211–227.

Freud, S. (1900). The interpretation of dreams. In J. Strachey (Ed. & Trans.), *The standard*

[9]This is a slightly revised version of an earlier paper (Rubin, 1985).

edition of the complete psychological works of Sigmund Freud, Vols. 4 & 5. London: Hogarth Press.

Freud, S. (1904). Freud's psychoanalytic procedure. In J. Strachey (Ed. & Trans.), *The standard edition of the complete psychological works of Sigmund Freud*, Vol. 7 (pp. 249–254). London: Hogarth Press.

Freud, S. (1911). Formulations on the two principles of mental functioning. In J. Strachey (Ed. & Trans.), *The standard edition of the complete psychological works of Sigmund Freud*, Vol. 12 (pp. 218–226). London: Hogarth Press.

Freud, S. (1912). Recommendations to physicians practicing psycho-analysis. In J. Strachey (Ed. & Trans.), *The standard edition of the complete psychological works of Sigmund Freud*, Vol. 12 (pp. 111–120). London: Hogarth Press.

Freud, S. (1924). The resistances to psychoanalysis. In J. Strachey (Ed. & Trans.), *The standard edition of the complete psychological works of Sigmund Freud*, Vol. 19 (pp. 213–222). London: Hogarth Press.

Freud, S. (1937). Analysis terminable and interminable. In J. Strachey (Ed. & Trans.), *The standard edition of the complete psychological works of Sigmund Freud*, Vol. 23 (pp. 209–254). London: Hogarth Press.

Glover, E. (1955). *The technique of psychoanalysis*. New York: International Universities Press.

Goldstein, J. (1976). *The experience of insight: A natural unfolding*. Santa Cruz, CA: Unity Press.

Goleman, D. (1976). Meditation and consciousness. *American Journal of Psychotherapy*, *30*, 41–54.

Goleman, D. (1977). *The varieties of the meditative experience*. New York: Dutton.

Gray, P. (1973). Psychoanalytic technique and the ego's capacity for viewing intrapsychic activity. *Journal of the American Psychoanalytic Association*, *21*, 474–494.

Greenson, R. (1967). *The technique and practice of psychoanalysis*. New York: International Universities Press.

Grotjahn, M. (1950). About the *"third ear"* in psychoanalysis: A review and critical evaluation of Theodor Reik's *Listening with the third ear: The inner experience of a psychoanalyst. Psychoanalytic Review*, *37*, 56–65.

Gyatso, T. (1982). You have every right to investigate religions. *Reality*, *1*, 1–3.

Heimann, P. (1977). Further observations on the analyst's cognitive process. *Journal of the American Psychoanalytic Association*, *25*, 313–333.

James, W. (1962). *Psychology: Briefer course*. New York: Collier Books.

Khan, M. (1977). On lying fallow: An aspect of leisure. *International Journal of Psychoanalytic Psychotherapy*, *6*, 397–402.

Kornfield, J. (1977). *Living Buddhist masters*. Santa Cruz, CA: Unity Press.

Kornfield, J. (1979). Intensive insight meditation: A phenomenological study. *Journal of Transpersonal Psychology*, *11*, 41–58.

Kornfield, J. (1981, October). The seven factors of Enlightenment. *The Ten Directions*, 1–4.

Langs, R. (1978). *The listening process*. New York: Jason Aronson.

Levenson, E. (1972). *The fallacy of understanding*. New York: Basic Books.

Little, M. (1951). Counter-transference and the patient's response to it. *International Journal of Psycho-Analysis*, *32*, 32–40.

Mahasi, S. (1973). *The progress of insight: A modern Pali treatise on Buddhist Satipatthana meditation*. Kandy, Sri Lanka: Buddhist Publication Society.

Mahasi, S. (1978). *Practical Vipassana meditational exercises*. Rangoon, Burma: Buddhasāsanānuggaha Association.

Marzetta, B.R., Benson, H., & Wallace, R.K. (1972). Combatting drug dependency in young people: A new approach. *Medical Counterpoint, 4*, 13–36.

Masson, J.M., & Masson, T.C. (1978). Buried memories on the Acropolis: Freud's response to mysticism and anti-Semitism. *International Journal of Psychoanalysis, 59*, 199–208.

McLaughlin, J.T. (1975). The sleepy analyst: Some observations on states of consciousness in the analyst at work. *Journal of the American Psychoanalytic Association, 23*, 363–382.

Nyanaponika, T. (1962). *The heart of Buddhist meditation*. New York: Samuel Weiser.

Nyanaponika, T. (1972). *The power of mindfulness*. San Francisco: Unity Press.

Olinick, S. (1980). *The psychotherapeutic instrument*. New York: Jason Aronson.

Oppenheimer, R. (1954). *Science and the common understanding*. New York: Simon & Schuster.

Reich, A. (1966). Empathy and countertransference. In *Psychoanalytic contributions* (pp. 344–360). New York: International Universities Press.

Rubin, J.B. (1985). Meditation and psychoanalytic listening. *Psychoanalytic Review, 72*(4), 599–612.

Schafer, R. (1976). *A new language for psychoanalysis*. New Haven, CT: Yale University Press.

Searles, H. (1959). The effort to drive the other person crazy: An element in the aetiology and psychotherapy of schizophrenia. *British Journal of Medical Psychology, 32*, 1–18.

Seng-tsan. (1984). *Verses on the faith mind* (R.B. Clarke, Trans.). Fredonia, NY: White Pine Press.

Shafii, M. (1973). Silence in the service of ego: Psychoanalytic study of meditation. *International Journal of Psycho-Analysis, 54*, 431–442.

Sharpe, E.F. (1930). The technique of psycho-analysis. In *Collected Papers on Psychoanalysis* (pp. 9–122). London: Hogarth Press.

Sterba, R. (1934). The fate of the ego in analytic therapy. *International Journal of Psycho-Analysis, 15*, 117–126.

Stolorow, R., & Lachmann, F. (1980). *The psychoanalysis of developmental arrests: Theory and treatment*. New York: International Universities Press.

Suzuki, S. (1970). *Zen mind, beginner's mind*. New York: Weatherhill.

Thera, S. (1941). *The way of mindfulness: The Satipatthana Sutra and commentary*. Kandy, Sri Lanka: Buddhist Publication Society.

Tulku, T. (Ed.). (1975). *Reflections of mind: Western psychology meets Tibetan Buddhism*. Emeryville, CA: Dharma Publishing.

Walsh, R. (1980). The consciousness disciplines and the behavioral sciences: Questions of comparison and contrast. *American Journal of Psychiatry, 137*, 663–673.

Welwood, J. (Ed.). (1979). *The meeting of the ways: Explorations in East/West psychology*. New York: Schocken Books.

Wilber, K. (1980). *The Atman Project: A transpersonal view of human development*. Wheaton, IL: The Theosophical Publishing House.

Wortz, E. (1982). Application of awareness methods in psychotherapy. *Journal of Transpersonal Psychology, 14*, 61–68.

7

On Resistance to Meditation
Psychoanalytic Perspectives

Discouraged by his inconsistent commitment to meditation practice, Brian consulted a teacher of meditation. He informed the teacher that he had greatly benefitted from meditating; he continually experienced it as a remarkable tool for cultivating awareness, insight, and compassion, yet he frequently did not do it. He wondered why he neglected something that he valued. Was he too busy? Lazy? Gripped by a deeper conflict?

The teacher informed him that he was afflicted by "sloth and torpor," namely, sluggishness and laziness, one of the five classical "hindrances" to Buddhist meditation (Narada, 1975). Brian was encouraged to pursue his meditation practice with greater diligence. For several weeks he dutifully attempted to carry out these instructions. To his dismay, sloth and torpor did not disappear or recede despite his earnest efforts to follow the teacher's advice.

In the hope of curing his malady, he examined some of the standard Buddhist texts on meditation practice (Buddhaghosa, 1976; Narada, 1975), which focused on conscious emotional and situational interferences to meditation, such as egoism, anger, sense desire, restlessness, blind adherence to rites and rituals, and theoretical studies divorced from practice, only to find the same diagnosis and prescription. He began to wonder if there could be other explanations for his ambivalent relationship to meditation. Does the classical Buddhist account offer a comprehensive explanation of the interferences to meditation practice? Is there a more illuminating way of understanding blockages to meditation?

Brian subsequently consulted a psychoanalyst. What gradually emerged in the course of therapy was Brian's feeling that his life had rarely been his own. His parents viewed him as an extension of themselves whose responsibility was to fulfill their goals for him and thus raise their impaired sense of self-worth. They demanded that he be what they wanted him to be. Rebellion was his only place of autonomy. In the course of therapy, he learned that he experienced participating in activities such as meditation as being the slave of his parent's expectations. His "laziness" about meditating was an attempt to avert being controlled yet again. Avoiding meditation was thus a way of asserting his freedom.

Almost 2500 years have elapsed since Buddha described the manifestations of what he termed the hindrances, impediments, and fetters to meditation practice. Learning to skillfully manage resistance is a crucial aspect of Buddhist practice. And yet, despite the fact that resistance plays an indispensable role in the meditative process, it has been neglected and incompletely explicated in the meditative literature. To be sure, the classical (Narada, 1975; Buddhaghosa, 1976) and contemporary Buddhist accounts (Goldstein, 1976; Goleman, 1977; Kornfield, 1977; Levine, 1979; Walsh, 1981) have outlined many of the conscious personal and environmental interferences to meditation practice: the six "hindrances," such as sense desire, anger, restlessness, sloth and doubt (Narada, 1975); the ten "impediments" (Buddhaghosa, 1976), e.g., excessive involvement with projects and theoretical studies divorced from practice (Buddhaghosa, 1976; Goleman, 1977), and the "ten fetters," including attachment to blissful nonordinary states of consciousness, adherence to (wrongful) rites and ceremonies, ignorance, and self-centered thinking (Narada, 1975). The unconscious psychological obstacles to practice, however, have not been systematically elaborated.

It remains essential for us, as it was for Buddha, to understand these persistent adverse interferences to meditation. Have Buddhist accounts exhausted the possibilities? Or, are there alternative explanations that could supplement and enrich Buddhist perspectives?

Reconsideration of the issues of interferences to meditation practice is thus timely. In this chapter, I will (1) review some of the important Buddhist contributions to an understanding of interferences to meditation, and (2) utilize psychoanalytic explanations of interferences to the psychoanalytic process (resistances and defense mechanisms) to illuminate interferences to meditation practice.

I shall approach psychoanalysis and Buddhism in a way that has not, to my knowledge, been done. I will illustrate how psychoanalysis, an exquisite cartography of strategies of self-deception and self-inhibition, can illuminate a crucial and neglected aspect of Buddhism: what interferes with developing, maintaining, and deepening meditation practice. Sound practice involves both knowing how to meditate and then actually doing it. The former has been amply documented (Goldstein, 1976; Kornfield, 1977; Mahasi, 1973; Nyanaponika, 1962, 1972); the latter requires further elaboration.

My argument will be that the highly elaborated psychoanalytic account of resistance and defensive processes (especially conflicts involving superego and ego ideals, transferences to authority figures, and traditional defensive processes, particularly denial, repression, inhibition, rationalization, intellectualization, undoing, and turning against the self, the major interferences to the psychoanalytic process) offers a more comprehensive delineation of the unconscious interferences to meditation than the traditional Buddhist explanation of the hindrances (Narada, 1975), impediments (Buddhaghosa, 1976), and fetters (Narada, 1975), and thus supplements and enriches the incisive, though incomplete, Buddhist discussion. This knowledge can, I believe, facilitate an improved capacity to manage difficulties with meditation practice.

In the first section, I will delineate the Buddhist perspective on the hindrances, fetters, and impediments. Then I shall discuss those defense mechanisms and aspects of resistance that can contribute to an understanding of the obstacles to meditation practice. In the concluding section some suggestions will be offered concerning how psychoanalytic perspectives might be used by teachers and practitioners of meditation as a tool in working with impediments to practice. Believing that theorizing cannot avoid creating models that simplify a more complex reality, my recommendations in the final section are offered in the spirit of provisional suggestions rather than conclusive pronouncements. The error, from this perspective, is not in proposing models but literalizing and reifying them and thus losing sight of their provisional nature. The data that will be utilized to illustrate the theoretical claims will be based on clinical cases, personal experiences, and the experience of friends and other meditators. In several cases I will utilize composite portraits.

Freud's (1912) seminal insight about the centrality of resistance in the psychoanalytic process, namely, that "the resistance accompanies the treatment step by step" (p. 103), applies, with the appropriate changes, to meditation practice. In psychoanalysis the process of free association produces resistance. In cultivating free association and thus facilitating the emergence of painful thoughts, feelings, and fantasies and challenging the way meditators view themselves and the world, meditation is potentially subversive of our mode of living. Hence the resistance to it.

It is the rare meditator who can practice effortlessly. Resistance to meditation—one "forgets" to meditate or chronically avoids it; falls asleep while meditating; "does not have enough time to do it"; uses it to suppress or escape from pain or conflict—frequently disrupts or threatens the very existence of one's practice.

HINDRANCES

Sense desire is the wish for, or the grasping after pleasant experiences and or sense objects. For example, Ralph precipitously interrupts his morning meditation to get something to eat after fantasizing about food while meditating. He is caught in sense desire.

Hatred or anger signifies states of ill will, aversion, annoyance, and irritation directed toward oneself or others. In the Buddhist literature these states of mind have been compared to "picking up a burning ember in our bare hands" (Levine, 1979, p. 79). Bill becomes so consumed with anger while meditating that he is unable to continue.

Sloth and torpor create a state of mental sluggishness, inactivity, and laziness. Often Barbara would not meditate because she felt too tired and lethargic.

Restlessness and brooding reflect mental disquietude. Harold often interrupted his meditation because he felt too agitated to continue.

Doubts involve questions and uncertainty about such issues as one's capacity to meditate and the validity of meditational theory and practice. That happened to me for a time after hearing the Indian philosopher Jiddu Krishnamurti's warning that meditation could become a form of psychic numbing and self-imposed slavery. A simile illustrates the effects of these hindrances:

> Imagine a pond of clear water. Sense desire is like the water becoming colored with pretty dyes. We become entranced with the beauty and intricacy of the color and so do not penetrate to the depths. Anger, ill will, aversion, is like boiling water. Water that is boiling is very turbulent. You can't see to the bottom.... Sloth and torpor is the pond of water covered with algae, very dense. One cannot possibly penetrate to the bottom because you can't see through the algae.... Restlessness and worry are like a pond when wind-swept. The surface is agitated by strong winds.... Doubt is like the water when muddied; wisdom is obscured by murkiness and cloudiness. (Goldstein, 1976, p. 53)

THE TEN FETTERS

Because five of the fetters are identical to the hindrances, I will discuss only those that are different: false view of the self, adherence to wrongful rites and ceremonies, attachment to realms of form, and conceit.

False view of the self is a difficult and widely misunderstood notion. It is often badly translated as nonexistence or no-self. There are two kinds of false views: (1) belief in the survival after death of a separate, immortal, unchanging soul or self eternally distinct from the body, or (2) belief in a soul or self that is identical with the body or the body-plus-mind that perishes at physical death (cf. Engler, 1983). Underlying both beliefs is the notion that "I" am an enduring and unchanging entity or "self" existing over time: the thinker of thoughts, doer of deeds, the agent of action. There are two ways that this fetter might manifest itself as an obstacle in meditation practice. A student of meditation might not feel a sense of urgency about cultivating mindfulness because immortality allows for infinite time to cultivate his or her meditation practice. One does not have to do anything now because there will always be time to do it later. False view of the self is illustrated by the student whose preoccupation with "dreams about a bright meditative future" (Johansson, 1983) diverts attention from carefully examining the working of the mind in the present.

Adherence to wrongful rites and ceremonies refers to the erroneous view that any external observance, whether religious ritual or rules of socially sanctioned conduct, is in and of itself sufficient for liberation. Ellen spent all of her time practicing religious rituals such as chanting Sanskrit verses rather than meditating.

Attachment to realms of form is the desire to hold onto or repeat blissful or ecstatic nonordinary states of consciousness. After some blissful moments meditating, Tom became more concerned with reexperiencing these states of consciousness than with cultivating awareness. He neglected meditation after discovering that these experiences occurred infrequently for him.

Conceit refers to the entire range of self-centered thinking from the ordinary sense in which consciousness is dominated by the awareness of I and mine in its most conspicuous forms to subtler manifestations that are prevalent until Enlightenment (Engler, 1983). Susan frequently compared herself to other meditators and felt diminished, and then did not meditate, whenever her progress did not measure up to her standards.

IMPEDIMENTS

The Visuddhimagga, a classical Buddhist examination of the theory and practice of meditation, lists ten external or situational "impediments" to meditation: (1) worrying about the maintenance of one's dwelling; (2) concern about one's family; (3) obtaining gifts or reputation that result in spending time with admirers; (4) being occupied with students or teaching; (5) involvement in projects; (6) travel; (7) caring for family or co-residents, pupils, preceptors, teachers, or fellow students in a monastery; (8) any kind of affliction or illness; (9) theoretical studies that detract from one's meditation practice; and (10) attachment to supernormal powers rather than involvement in meditation.

Contemporary Buddhist teachers suggest four additional obstacles to meditation. Many meditators, especially in the West, often engage in "self-psychotherapy" rather than meditation. Cultivating meditative concentration is essential to deepening one's meditation practice. Preoccupation with specific mental content, such as emotional patterns, historical antecedents, and dynamic meaning, interferes with the training of concentration and the development of insight into more subtle aspects of mental processes.

A related barrier to meditation among Western students is their lack of faith in the meditative process (Walsh, 1981). Buddhism distinguishes between blind and intelligent faith. Students of meditation are encouraged to sail a balanced course between paralyzing skepticism and unquestioning allegiance. Faith can be a motivational fuel that allows one to be patient and persevere.

"Right effort" is one of the essential ingredients of the meditative process. Lack of faith impedes intensity of effort. Asian teachers of meditation have remarked that Western students of meditation have trouble with effort; they do not truly appreciate the intensity of effort required in meditation and are less inured to discomfort than Asian practitioners. They tend to have trouble tolerating discomfort (Walsh, 1981).

WHAT PSYCHOANALYSIS OFFERS BUDDHISM

My experience as a psychoanalyst and meditator convinces me that Buddhist explanations of resistance provide a necessary but incomplete account of what interferes with meditation practice. The problem, I believe, is this: Interferences to meditation are acknowledged although not fully clarified in Buddhism. A classical Buddhist text dealing with eight ways people relate to frustration is illustrative. In "eight recalcitrant men and their eight defects" (Johansson, 1983, p. 19), several defensive processes, e.g., forgetfulness, aggression, projection, denial, and withdrawal are described but not explained. Why people utilize these strategies is not clarified and remains a mystery. The case of Jody taught me that the limitations of such a perspective are not trivial. Jody suddenly expressed a lack of interest in meditation. From a Buddhist perspective, she was afflicted by the hindrance of laziness.

Jody was the youngest of two sisters. Her older sister was considered a "saint" and favored by her parents. Exploration in therapy revealed that Jody was uncomfortable with the positive changes that resulted from meditation, especially her increased self-assertion. Her family complained that she was no longer the "sweet" person she used to be. In adolescence Jody had developed a strategy of denying her own needs and becoming more "saintly" than her older sister in order to be accepted by her parents. She became inhibited about meditating because it threatened the saintly role she was so invested in cultivating. Focusing on her manifest behavior—her apparent sloth—rather than illuminating its unconscious meaning would have been a flawed strategy for alleviating the resistance.

The hallmark of psychoanalytic investigations of the obstacles to the psychoanalytic process are the concepts of "resistance" and "defense mechanisms." Resistance refers to those factors that impede the treatment process. "Whatever interrupts the progress of analytic work," notes Freud (1900), "is a resistance" (p. 517). Successful treatment hinges on effective resolution of resistance. Failure to understand resistance can undermine treatment.

The psychoanalytic literature on resistance since Freud's groundbreaking discoveries has been somewhat scattered. There is no organizing model or single frame of reference for understanding resistance (Milman & Goldman, 1987).

I shall use resistance and defense in a more inclusive and, I feel, more heuristic way than much of traditional psychoanalytic usage. Resistance, in my view, and in the view of many contemporary analysts, should be used to discuss the avoidance of any thought, feeling, or fantasy that is upsetting or threatening. In the spirit of several recent revisions of psychoanalytic theorizing (Kohut, 1984; Stolorow & Atwood, 1984), I maintain that it is advisable to think of defensive processes or activities rather than mechanisms. Defensive processes as a major mode of resistance refer to the specific strategies employed in avoiding or warding off frightening or painful thoughts, feelings, or fantasies, and thus safeguarding one from anticipated danger, vulnerability, pain, and suffering and in extreme cases preserving one's sense of psychological coherence, continuity, and survival (Kohut, 1984).

The difference between resistance and defense is illustrated by one meditator I know who chooses to teach classes on Asian psychology rather than meditate because distasteful feelings arise whenever she meditates. Avoiding meditation is the resistance, and intellectualization and rationalization are the defensive processes.

Meditation may be resisted because it is a threat to one's style of adaption (Sandler, Dare, & Holder, 1973). There was a period in the practice of Anne, a teacher of meditation from an alcoholic and sexually abusive family, when she avoided meditation because it fostered the emergence of excruciatingly painful repressed memories and feelings. Staying away from meditation was an attempt to ward off experiences that threatened the psychological balance that she had previously maintained (Kornfield, 1987, personal communication).

Unconscious guilt or one's need for punishment (Freud, 1926) can lead to resistances to meditation. Andrea did not meditate on a regular basis because it made her happy, which clashed with her need for punishment and self-deprivation (Carrington, 1978).

Transference resistances (Freud, 1926) can also cause interferences to meditation. The student transfers significant aspects of his or her infantile past, especially feelings about and ways of relating to important early figures onto the teacher. An example would be a meditation student with erotic transference wishes toward the meditation teacher. She experiences the teacher's relative neutrality as a rebuff and stops meditating in order to punish the teacher.

Kohut's (1977, 1984) pioneering delineation of "mirroring" and "idealizing" transferences illuminates two other potential resistances to meditation. A student in an idealizing transference, that is, needing the teacher to be the kind of all-powerful parental figure she or he so sorely lacked in childhood, replaces the healthy skepticism necessary to meditative development with blind faith toward a deified teacher. One meditator in a mirroring transference with his Buddhist teacher, by which I mean he needed and wished to be deeply valued, had great difficulty maintaining continuity in his meditation practice after his teacher treated him respectfully but without enthusiasm.

Difficulties in external relationships can also interfere with meditation practice. Alice consulted a therapist because of blockages in her meditation practice. The therapist learned that Alice's husband was having an affair with another woman. Alice was unaware of any negative feelings toward her husband until she meditated. While meditating she became faintly aware of anger and grief. The increased self-awareness that meditation facilitated threatened to further disrupt her relationship with her husband (Kornfield, 1987, personal communication).

Meditation is sometimes avoided because of the threat one associates with success (Freud, 1926). For example, one meditator associated consistent meditation practice with success and success with sibling envy. Interrupting his practice was a way of avoiding success and the anticipated guilt he strove not to experience.

Meditation is sometimes resisted because of what Abraham (1919) termed the threat to one's self-esteem. Larry only felt good about himself when he lived up to an idealized image of how he should be. He resisted meditation because it revealed that he actually felt anger, jealousy, and vanity, which in turn made him feel humiliated and mortified.

In resistance based on the gain from illness (Freud, 1926), benefits are obtained from being ill, such as being pitied or taken care of or exacting revenge against others who share in the patient's suffering. I can imagine the case of a meditator who unconsciously sabotages his meditation practice in order to punish a teacher who he believes does not appreciate him.

Meditation may be avoided so as to maintain or restore precarious self or object images or ward off configurations of experience that are felt to be conflictual or dangerous (Stolorow & Lachmann, 1984–5). It is in this context that recent descriptions of various meditation-induced disturbances, such as delusions, paranoia, depersonalization, and derealization (Epstein & Lieff, 1981), may in part be understood. Meditators, like analysands, attempt to avoid or decrease the possibility of experiencing such symptoms. Jan, a meditator with a fragile sense of self, illustrated this type of interference to meditation. She needed constant attention from others because of her depleted sense of self. She avoided meditation because being alone with herself while meditating caused her to experience a dreaded sense of isolation and terror.

PSYCHOANALYTIC DEFENSE MECHANISMS

The specific strategies one employs in resisting (defensive processes) are unconsciously designed to eliminate or decrease awareness of distressing thoughts, feelings, or fantasies. One might not meditate so as to avoid anxiety, guilt, longing, grief, shame, envy, rage, fear, or humiliation.

There is no universally accepted list of defense mechanisms (White & Gilliland, 1975). Any list may be incomplete and open to criticism since there is a difference of opinion among psychoanalysts about what should be termed a defense mechanism (Brenner, 1982). A composite list might include denial, repression, conversion, inhibition, negation, condensation, displacement, rationalization, intellectualization, reaction formation, undoing, isolation of affect, regression, projection, turning against oneself, dissociation, splitting, identification with the aggressor, turning active into passive, projective identification, and sublimation. The defensive processes that seem especially applicable to resistance in meditation are denial, repression, rationalization, undoing, turning against oneself, and intellectualization.

Denial is the process of excluding from awareness and distorting the actual nature of some disturbing experience (Brenner, 1974, 1982). One meditator I knew who had an emotionally depriving childhood denied both the positive effects of meditating and the negative consequences of avoiding it. Denial prevented the awareness of unconscious guilt over being emotionally nourished and fulfilled, states to which he did not feel entitled.

Repression occurs when one excludes from awareness some painful aspect of reality and then remains unaware of the exclusion (Freud, 1926). A meditator repeatedly forgot that he had not meditated and remained unaware of forgetting. Remembering might engender guilt because he had failed to live up to his image of being a dedicated meditator.

Rationalization refers to logical and believable explanations for avoiding a course of action. A meditation teacher I know shies away from meditating and focuses instead on teaching others about Asian psychology. Teaching is less threatening to her than meditating, which increases her awareness of distasteful feelings of dependency and inadequacy.

Undoing involves actions that undercut or cancel out prior thoughts or acts. Barbara was one of eleven children. When she was ill as a child, her overworked mother removed herself emotionally from family affairs. Barbara felt abandoned. She stopped meditating when it was no longer experienced as a panacea and she then felt that it was "indifferent" to her and had "deserted" her just when she needed support (Carrington, 1978).

The defensive process of turning against oneself involves directing feelings such as anger against oneself rather than toward the person with whom one is upset. An example of this would be a meditator who was enraged with his boss. Instead of telling the boss how he felt, which he feared might threaten their relationship and his job security, he expressed the feelings toward himself by neglecting formerly meaningful activities such as meditation.

The Buddhist and psychoanalytic processes that I have discussed illuminate what interferes with greater involvement in meditation practice. Defensive processes, e.g., denial and intellectualization, can elucidate a different and neglected barrier to deepening meditation practice:

The ways that it is pursued that are problematic. The ideal in Buddhism is a balanced practice (Kornfield, 1987, personal communication). Normally resistance is discussed from a perspective of how one avoids meditation; but how it is approached can be problematic, leading to blindspots and imbalances. This idea is not foreign to Buddhism, but it has not been fully elaborated. In the Visuddhimagga (Buddhaghosa, 1976), the seductive nature of meditative ambition is mentioned but not elucidated. During the stage of practice called the "Knowledge of Arising and Passing Away" (Goleman, 1977, p. 28), meditators may experience the "ten corruptions of insight": visions of a bright light, rapturous feelings, bodily or mental tranquility, devotional feeling toward Buddhist teachings, Buddha, and the teacher, a sense of vigor in meditating, sublime happiness and bliss, accurate and clear perception of each moment of experience, effortlessness and mindfulness, mental equanimity, and a subtle attachment to the aforementioned experiences (Goleman, 1977, p. 109). The danger is in becoming attached to these experiences, and thus "mistaking what is not the Path for the Path" (Goleman, 1977, p. 28). Like a mountain climber who does not reach her destination because she gets sidetracked along the way, attachment to these experiences impedes deepening of meditative insight and progress.

Psychoanalytic perspectives on defensive processes, e.g., denial and intellectualization, can reveal pitfalls in the way meditation is approached. Denial is evident in cases in which meditators concentrate the mind and thus emotionally anesthetize themselves from psychological pain. One female meditator from a traumatic childhood background hated her body and had a series of failing relationships with men. Her lesbian relationship clashed with her homophobic ideals. By deeply concentrating her mind in meditation, she blocked out these struggles (Kornfield, 1987, personal communication). She used meditation to suppress rather than resolve her conflicts.

The meditative technique taught in one of the core lineages of contemporary Theravadin Buddhism, which involves noting and labeling each moment of experience (Mahasi, 1978), can be utilized by some students as a form of intellectualization. Proficiency in categorizing "loaded" affective states and experiences replaces direct contact with them.

PSYCHOANALYTIC PERSPECTIVES ON WORKING WITH RESISTANCE IN MEDITATION PRACTICE

Meditators, like analysands, neglect resistance at their own peril. Unanalyzed resistances do not usually disappear and can lead to blindspots and blockages in one's meditation practice. In severe cases one's practice can even be undermined.

Psychoanalysis is distinguished from all other therapeutic and soteriological systems by its thorough and systematic analysis of resistance. It provides the most comprehensive and sophisticated examination of resistance of any spiritual or psychological tradition that I am aware of.

Resistance analysis, like a heart transplant, is a complex process that requires specific conditions—the "analytic situation"—for its success. Because the meditative situation lacks some of the necessary conditions to be a congenial host in that there is no provision for examining the vicissitudes of the student's psychic life (that is, there is no therapeutic relationship or alliance devoted to illuminating the student's way of relating or being), psychoanalytic strategies for working with resistance in meditation cannot be applied without significant modification. And yet, despite these difficulties, psychoanalytic perspectives can assist meditation teachers in working with resistance.

The first step in working with resistance is to recognize its presence. Effective resolution cannot occur unless resistance is recognized. This is not so difficult when the resistance is obvious, as when one does not want to meditate. It is more complex when the resistance is more subtle, e.g., one uses meditation to anesthetize one's emotional pain. Recognition of resistance is complicated by the fact that there may be a resistance to noticing that one is resisting. One student who felt that his parents only conditionally loved him had grandiose fantasies of being a world-renowned teacher of meditation. He had great difficulty examining his resistance to meditation because it felt like a "defect" that he feared would lead to loss of love.

The teacher's empathy, intuition, theoretical knowledge of human nature and the meditative process, and sensitivity to his or her fantasies, images, and induced reactions to the student may be valuable in detecting subtle interferences to meditation. For example, I once became aware of the resistance underlying a student's apparent interest in discovering what was interfering with his meditation practice by paying attention to subliminal images of a "tug of war" I had during his description in a session of his "wish" to figure out why he was ambivalent about meditating.

Military metaphors are frequently used in psychological literature on resistance. It is counterproductive to view resistance in an adversarial way (Kohut, 1984). It is important to communicate to students of meditation that resistance is not indicative of emotional weakness, but is an inevitable aspect of the meditative process. An awareness that resistance is a strategy that one employs to protect against anticipated pain or danger can encourage a nonjudgmental and empathic stance.

The particular details of resistance must be demonstrated to the student. Two factors shape the teachers ability to do this: how hidden the resistance is and how willing the student is to examine it. One meditator I know has difficulty facing interferences in her practice because she feels diminished whenever she is not perfect. Allowing the resistance to develop may increase the likelihood that the teacher can demonstrate it to a student. A teacher may, for example, wait for several examples before mentioning it to a student. Resistance should be pointed out in a tactful and nonjudgmental manner.

The forms of resistance are virtually limitless (Sandler et al., 1973). It is necessary to clarify the motives and modes of resistance. The teacher needs to understand what painful affect or anticipated danger is being avoided and what means the student is utilizing to do so.

In order to do this, resistance should not be taken at face value (Carrington, 1978). Take the case of Stewart, a young man from a dysfunctional family who stopped meditating because he felt "agitated." He appears to be afflicted by the hindrance of restlessness and brooding. In therapy it emerged that he was scared that his meditation practice was going well. He unconsciously equated success with independence and aloneness. Because of his unconscious identification with and fear of separating from his emotionally disabled family, he dreaded success, and thus sought to avoid meditation.

Clarifying the resistance entails understanding how it works and functions. One needs to identify the subjective danger or emotional conflict that makes the resistance seem like an absolute emotional necessity (Stolorow & Lachmann, 1984–5). Teachers and students can ask themselves: What is the resistance in the service of? What is it directed against? What is the student attempting to accomplish by resisting? Are there any important relationships with others, actual or internalized, that might be threatened if one meditated?

In encouraging a removal of self-censorship and a freedom to move effortlessly and spontaneously from one thought or image to another (Lothane, 1987, personal communication), free association is an invaluable tool in elucidating unconscious determinants of resistance. Teachers can encourage student's to "say what comes to mind" about the resistance. Outside of a retreat context, students could have a dialogue with the resistance, paint it, or physically embody it.

Once the resistance is clarified, it must be interpreted. To interpret means to make an unconscious phenomenon conscious; to elucidate the unconscious meaning and causes of a psychic event (Greenson, 1967).

Interpretation involves pointing out that one is resisting, how one is resisting, and why one is resisting. The language of an interpretation should be clear, concrete, and direct. Teachers should be sensitive to timing, tact, and tone.

The previous stages of resistance analysis pave the way for elaborating the working-through phase of interpretations of the resistance, leading to lasting change in reaction or behavior (Greenson, 1967). Here one repeats and extends the earlier processes until there is an integration of insight and behavior. If a Buddhism of the future utilizes psychoanalytic understandings of resistance and defensive processes, then meditator's might have a greatly enhanced capacity to cope with difficulties in meditation practice.

REFERENCES

Abraham, K. (1919). A particular form of neurotic resistance against the psychoanalytic method. In K. Abraham, *Selected papers on psycho-analysis* (pp. 303–311). London: Hogarth Press, 1927.

Brenner, C. (1974). *An elementary textbook of psychoanalysis.* New York: Doubleday Anchor Books.

Brenner, C. (1982). *The mind in conflict.* New York: International Universities Press.

Buddhaghosa, B. (1976). *The path of purification* (B. Nyanamoli, Trans.). Berkeley, CA: Shambhala.

Carrington, P. (1978). *Freedom in meditation.* New York: Doubleday Anchor Books.

Engler, J. (1983). Vicissitudes of the self according to psychoanalysis and Buddhism: A spectrum model of object relations development. *Psychoanalysis and Contemporary Thought, 6*(1), 29–72.

Epstein, M., & Lieff, J. (1981). Psychiatric complications of meditation practice. *Journal of Transpersonal Psychology, 13*(2), 137–147.

Freud, S. (1900). The interpretation of dreams. In J. Strachey (Ed. & Trans.), *The standard edition of the complete psychological works of Sigmund Freud*, Vols. 4 & 5. London: Hogarth Press.

Freud, S. (1912). Recommendations to physicians practicing psycho-analysis. In J. Strachey (Ed. & Trans.), *The standard edition of the complete psychological works of Sigmund Freud*, Vol. 12 (pp. 111–120). London: Hogarth Press.

Freud, S. (1926). Inhibitions, symptoms and anxiety. In J. Strachey (Ed. & Trans.), *The standard edition of the complete psychological works of Sigmund Freud*, Vol. 20 (pp. 77–175). London: Hogarth Press.

Goldstein, J. (1976). *The experience of insight: A natural unfolding.* Santa Cruz, CA: Unity Press.

Goleman, D. (1977). *The varieties of the meditative experience.* New York: Dutton.

Greenson, R. (1967). *The technique and practice of psychoanalysis.* New York: International Universities Press.

Johansson, R.E.A. (1983). Defense mechanisms according to psychoanalysis and the Pali Nikayas. In N. Katz (Ed.), *Buddhist and Western psychology* (pp. 11–24). Boulder, CO: Prajina Press.

Kohut, H. (1977). *The restoration of the self.* New York: International Universities Press.

Kohut, H. (1984). *How does analysis cure?* Chicago: University of Chicago Press.

Kornfield, J. (1977). *Living Buddhist masters.* Santa Cruz, CA: Unity Press.

Levine, S. (1979). *A gradual awakening.* New York: Doubleday Anchor Books.

Mahasi, S. (1973). *The progress of insight: A modern Pali treatise on Buddhist Satipatthana meditation.* Kandy, Sri Lanka: Buddhist Publication Society.

Mahasi, S. (1978). *Practical Vipassana meditational exercises.* Ragoon, Burma: Buddhasasananuggaha Association.

Milman, D.S, & Goldman, G. (1987). *Techniques of working with resistance.* Northvale, NJ: Jason Aronson.

Narada, T. (1975). *A manual of Abhidamma.* Colombo, Sri Lanka: Buddhist Publication Society.

Nyanaponika, T. (1962). *The heart of Buddhist meditation.* New York: Samuel Weiser.

Nyanaponika, T. (1972). *The power of mindfulness.* San Francisco: Unity Press.

Sandler, J., Dare, C., & Holder, A. (1973). *The patient and the analyst.* New York: International Universities Press.

Stolorow, R., & Atwood, G. (1984). *Structures of subjectivity: Explorations in psychoanalytic phenomenology.* Hillsdale, NJ: The Analytic Press.

Stolorow, R., & Lachmann, F. (1984–5). Transference: The future of an illusion. *Annual of Psychoanalysis, 12/13,* 19–37.

Walsh, R. (1981). Speedy Western minds slow slowly. *Revision, 4,* 75–77.

White, R.B., & Gilliland, R.M. (1975). *Elements of psycho-pathology—The mechanisms of defense.* New York: Guilford Press.

8

Spirituality and the Psychoanalyst

In this chapter, I shall explore what value spirituality in general and Buddhist psychology in particular might have for psychoanalysis. My focus will be on how Buddhism can enrich psychoanalysis, especially the experience of the analyst, not how psychoanalysis can complement Buddhism. One important and neglected consequence of the Eurocentrism that has pervaded psychoanalysis' relationship with Buddhism has been that psychoanalysis has neglected the resources of spirituality. It is thus not accidental or surprising that what spirituality might offer psychoanalysis has been insufficiently formulated.

After delineating some of the unique emotional and intellectual demands the practice of psychoanalysis and analytically oriented psychotherapies make on its participants, I shall discuss several ways that meditation, the core practice of Buddhism, can enrich psychoanalytic treatment and mitigate these strains. My remarks about psychoanalysis should be understood as also applying, where appropriate, to other forms of psychotherapeutic treatment.

The practice of psychoanalysis, and the analytically oriented therapies it has spawned, is a complex, unsettling, and arduous enterprise that makes intense emotional and intellectual demands on the analyst, who needs to decode unconscious communications, empathize with complex psychic states, understand intricate patterns of interpersonal interaction, and sometimes provide needed archaic emotional functions (cf. Kohut, 1977, 1984). By this I mean that the patient may demand that the analyst "mirror" or confirm and value the patient or the patient may insist that the therapist be an ideal figure that she or he could idealize, feel connected to, and draw sustenance from.

Concern with the mental health of the therapist is a neglected topic (Freudenberger, 1983). The dearth of literature on this issue seems symptomatic of the resistance in psychoanalysis to both exploring the inner experience of the psychoanalyst and acknowledging difficulties the psychoanalyst may have.

Practicing psychoanalysis or psychoanalytically oriented psycho-therapy is an inherently self-denying and self-negating role, since one forgoes and curtails many of the normal kinds of self-centeredness and self-gratifications that characterize most other human relationships, in-cluding those in the workplace. There is interpersonal contact but it is not mutual. The "relatedness" the analyst experiences is one-dimen-sional and skewed, involving understanding the client's experience and facilitating the therapeutic process (Freudenberger & Robbins, 1979) rather than attending to the experience and needs of the analyst. The analyst's own emotional life is a resource for understanding the client but generally is not a central focus of mutual concern.

The analyst spends most of her working day listening to and at-tempting to empathize with human experiences that may be distressing or even foreign. People come to analysts for a variety of reasons, ranging from wishing to resolve conflicts or enrich their lives to placating a spouse or parent or a member of the legal system. A common denomina-tor of the vast majority of the people who consult analysts is that they are in emotional pain and are psychologically needy. They often feel de-prived, depressed, entrapped, and powerless. With the daily fare of child abuse, incest, rage, envy, seductiveness, conflicts, excruciating confusion, alienation, loneliness and suffering, analysts may witness more psychological suffering in a week then the ordinary person might confront in a year.

Psychoanalyst Ralph Greenson (1966) notes two other strains on the clinician: the analyst "has to be free of the usual restricting conven-tionality of society and relatively indifferent to the superficialities of everyday life" (p. 276) and must "bear the roles the patient casts on him in his transference reactions, to endure being the hated enemy or rival, or the dearly beloved or the frightening father, or the seductive loving mother, etc." (p. 283). The analyst experiences a range of feelings, such as fear, sadness, intimacy, joy, jealousy, and pride, which his or her training and professional self-ideal discourage him or her from expres-sing. Whether the analyst is hated or loved, envied or idealized, he or she needs to remain persistent, attentive, benevolent, reliable, dedi-cated, and nonretaliatory. Periodic discouragements, disillusionments, nonrewarding stretches, doubts, and obstacles must be borne with at least a measure of equanimity.

There is a danger that therapists can be so preoccupied with the trials and tribulations of their clients that they neglect their own inner life, including symbolic representations (Freudenberger & Robbins, 1979, pp. 288–289). Neglecting these wellsprings of insight, nourishment, and creativity can lead to self-impoverishment.

The extraordinary isolation of the private practitioner may also exacerbate the self-abnegating and stressful conditions of being a therapist. Opportunities for social interchange and sharing that exist in most work situations are absent in the psychotherapeutic context. Some analysts spend whole days seeing no one but their patients and sometimes not even them! (cf. Cooper, 1986, p. 592). Repeated losses are built into the process. The analyst engages in intense and psychologically intimate relationships for months, years, or even decades that often end without further contact because of professional restrictions.

"Analysts," as Freud (1937) recognized, "are people who have learned to practise a particular art; alongside of this, they may be allowed to be human beings like anyone else" (p. 247). The human dimension is often omitted in psychotherapeutic training and in our view of ourselves and our conception of the therapeutic process. Stresses in the analyst's life, such as illnesses, deaths, births, family or relationship difficulties, distress over political or social realities, may also contribute to his or her depletion. Since the perfectly analyzed analyst is, in my view, an illusory ideal, personal issues that burden the analyst and the work are inevitable and may exacerbate the already existing complexities of analysis. These may include such things as unrealistic and personally burdensome expectations of and a sense of responsibility for the outcome of the treatment or the analyst's need to have the patient validate his or her worthiness or be just like him or her (a psychological twin). If one unconsciously entered the field in order to make reparation for unconscious guilt feelings deriving from the past, then he or she may, for example, experience emotional depletion when the analysand "fails" to validate the analyst's sense of specialness. If a therapist is attempting to compulsively rescue others as a way of disproving his or her conviction of inner badness, then treatments that are not progressing or are unfolding slowly may be a source of great frustration or even generate a sense of personal failure.

The nature of psychoanalytic training, which usually breeds submissiveness and self-debilitating perfectionistic ideals of self-mastery and control, often complicates therapeutic work and the capacity of therapists to address their difficulties. Psychotherapeutic training both insufficiently attends to and thus minimizes the hazards of therapeutic work and inculcates unrealistic ideals and expectations of psychic invulnerability.

The role strain of being placed as an adult, who may be in highly responsible positions in other facets of his or her life, including being a parent, partner in an intimate relationship, and so forth, in the inherently submissive stance of a student and beginning therapist is quite difficult. This is probably further compounded by being asked to perform the highly complex task of conducting analysis without highly developed theoretical or clinical tools. One important way that students and beginning therapists handle this tension is to comply with the demands of their training institution, hide vulnerability, bury theoretical disagreements, and inhibit risk and creativity. Fitting in with the institutional ethos, including minimizing self-vulnerability, enables trainees to solidify their precarious status. Embracing the theories of the school to which one identifies offers a sense of intellectual and emotional comfort. It serves as a map that guides and orients the analyst through the analytic process. It also gives one a stable identity and sense of belonging. But it fosters unrealistic ideals and expectations of self-knowledge, self-mastery, and selfless service, as well as a phobic stance toward emotional distress and vulnerability. Psychotherapists may thus have great difficulty acknowledging or coping with their own fallibility. Admitting difficulties or seeking help may be viewed as a stigma or an intolerable sign of weakness. One learns early on that it is politically and perhaps economically dangerous to admit weakness. Will colleagues refer as readily to a therapist who admits being in great turmoil?

Ironically, the shame psychotherapeutic training induces in many therapists when they struggle may cause them to respond to their difficulties with denial or secrecy. This inhibits both their ability to recognize inchoate signs of burnout and may interfere with their seeking assistance or support from colleagues or appropriate professionals. These difficulties cannot be addressed or remedied when they remain disavowed.

Psychotherapeutic training may unwittingly contribute to burnout in another way. By inhibiting intellectual risk, it may routinize therapeutic work, and thus contribute to burnout as one gradually becomes entrapped within the grip of an allegiance to Freud or the founder of the school of analysis or therapy to which one feels most affiliated. This can inhibit the emergence of one's innovative insights.

Because idol worship in the form of deification of one's analyst(s), supervisors, or the founder of the school of therapy with which one feels most identified is rarely worked through in most analyses (cf. Rubin, 1995), the therapist's own psychic individuation and authenticity, as embodied in the therapeutic risks and experiments he or she engages in, may be psychologically destabilizing. The analyst may experience debilitating guilt or shame over therapeutic interpretations or interventions that deviate from the therapeutic dictums of whatever figure(s) he or she idealizes.

The complex and alive experience of the therapeutic relationship and process is unwittingly converted, in such a situation, into what psychoanalyst Arnold Goldberg (1990) terms a "prison-house" of confining formulas and rules. Being intellectually straitjacketed can restrict one's capacity to understand and to freely and completely respond to the exigencies of the therapeutic process as one listens *for* confirmation of one's preferred theories rather than listening *to* the client's idiographic material. This impairs one's clinical effectiveness and often generates a sense of staleness and boredom. Burnout is likely in such a situation because it is difficult to feel effective and engaged when one is intellectually imprisoned within a restrictive theoretical framework.

Female analysts confront the aforementioned difficulties and stresses as well as additional burdens ranging from implicit and explicit sexism in society and the psychotherapeutic profession to the pressures of juggling work and family responsibilities, if they are part of a family. Sexism in the field emerges in the androcentric biases of psychoanalytic notions regarding human development and human subjectivity that treat experiences and values that are central to men, such as goals, individuation, and accomplishments, as normative for all humans (Rubin, 1995). This eclipses facets of life that women may also find important, such as non-self-centered states of being, relatedness, communion with nature, animals, and people, and just being.

Because women are culturally cast in the role of caregivers, with all the pressures that entails, the possibilities of burnout may be greater than for men. It is symptomatic of the problem that this topic has received scant attention.

Given the personal and institutional strains in the practice of psychoanalysis that I have enumerated, the need for concerted efforts to counteract the inevitable demands and stresses of being a therapist is evident, otherwise burnout is a possibility. Freud (1937) was aware of the inevitability of countertransference and the necessity for ongoing "mental hygiene" to help the analyst handle the difficulties of treatment. He recommended that analysts return to analysis every 5 years. In the next section, I shall attempt to paint a picture of how spirituality might offer an additional resource.

SPIRITUALITY AND THE CLINICIAN

Listening to ourselves and our analysands, as I suggested in Chapter 6, is both the essential tool of psychoanalytic inquiry and the foundation of psychoanalytic technique. Meditation deeply aids listening. It facilitates greater access to formerly unconscious material as well as greater receptivity to subtle mental and physical phenomena. One notices thoughts, feelings, fantasies, images, and bodily sensations that one is ordinarily unaware of. To cite one example among many: while involved in intensive meditation practice I have much greater access to and clarity about my own dream life including frequent occurrences of lucid dreaming.

Meditation practice also promotes greater tolerance for whatever we experience, including affect. Affect integration seems not to be promoted in many families. Many therapists were raised in familial contexts in which parents had difficulty helping them develop this capacity. When parents impede the development of this ability by doing such things as being disinterested in or stifling their child's expression of hurt or anger, panicking when their children are anxious, or prematurely and intrusively "solving" the child's problem, then affect integration may become disturbed or insufficiently developed in the child. If a therapist grew up in such an environment, then the intensity of feeling engendered in a particular treatment may pose a problem. Two typical strategies that clinicians I have treated or observed in supervision have employed are hiding behind a rigidly established set of guidelines for conducting the treatment so as to create predictability and detach from affect, and dulling emotional intensity in the session by utilizing intellectualized styles of organizing and interpreting analytic material.

Meditation, by developing the ability to open to the texture of experience with less attachment and aversion, aids the therapist in more skillfully handling affect. The analyst can literally sit with and through a greater range of affect without the need to shield him- or herself. This helps the therapist to bear the endemic losses and pain of therapy and the emotional distress that this generates more easily.

Analysts are not immune to critical thoughts, self-doubts about their ability to help certain patients, and countertransferential reactions. The affect tolerance that spiritually based disciplines such as meditation develop leads to decreased self-recriminative tendencies and greater self-acceptance.

A sense of nonattachment and inner equanimity organically develops from the greater tolerance and self-acceptance that meditation promotes. Therapists, like most clients, often unconsciously cling to whatever they desire and push away whatever they dislike; they react *from* what they are attached to, such as their frustration with a particular treatment, rather than *to* it. They then are more prone to ask, for example, "what have I (or the client) done wrong?" instead of "what is going on here?" Meditative practice cultivates a more impartial and nonreactive experiencing of both attractive and aversive phenomena. Since attachment to the degree of change or progress in the treatment can haunt the therapist or even in certain cases contribute to burnout, developing a stance of nonattached commitment to the treatment can be quite energizing and therapeutically useful.

By aiding the therapist in tolerating a wider range of experiences and reactions without fear or judgment, these experiences can be utilized as grist for the self-investigative mill. This opens up unexpected possibilities for learning and growth. Because our therapeutic work has self-negating facets, an analyst needs an opportunity for renewal and rejuvenation. The opportunity for self-revitalization that such learning promotes can lessen some of the depriving facets of being a therapist and make them more bearable.

The nonattachment that meditation practice develops also fosters greater freedom in the analyst. In cultivating perceptual acuity, attentiveness, and nonattachment, meditation fosters awareness of and deautomatization from previously habitual patterns (Deikman, 1982), including some of the unresolved issues from one's own analysis that create difficulty or conflict for the clinician in conducting therapy.

This freedom of responsiveness might lessen the submissiveness our training often promotes. Meditation practice has personally helped me adopt a more fluid relationship to the theories that I utilize to organize and make sense of the complex and overdetermined clinical actualities. It has helped me employ theoretical and clinical maps that I find illuminating, while simultaneously recognizing their ultimate provisionality and the inevitability of continually revising them.

The spiritual perspective that meditation practice promotes can also lessen the painful alienation that therapists experience. One of the hallmarks of Buddhism is the recognition of the inherent suffering that egoism and self-centeredness create. Buddhist practice, as I suggested in Chapter 3, fundamentally challenges the conventional, taken-for-granted Western conception of a unified, stable, autonomous self. It also promotes greater selflessness (Zvi Lothane, personal communication). What I termed "non-self-centric" moments of consciousness in Chapter 3 become more apparent when one meditates (cf. Rubin, 1993). In eroding restrictive and taken-for-granted self-identifications, our ordinarily narrow conceptions of ourselves, meditation practice can enhance self-experience by facilitating greater freedom, flexibility, and inclusiveness of self-structures.

Decreased self-absorption opens therapists up to the possibility of greater intimacy, for rapport and love demand that we unconstrict and sometimes transcend our normally more restrictive sense of separateness from others and the world. This often facilitates a heightened sense of living (Rubin, 1993). This does not eliminate the self-alienating and painful isolation of being a private practitioner, but it can decrease it by making us feel more connected to ourselves and others.

The capacity that spiritual practice engenders of *just being* can also be helpful in promoting less self-alienation in the therapist. Since the therapist needs to cope with "fullness" (an overabundance of feelings) as well as emptiness (a sense of depletion), nondoing, what is termed *wu wei* in Taoism, may be what some analysts need. This might entail building in more free time in one's life just to be or engaging in practices of self-care or self-renewal such as yoga, painting, playing music, martial arts, gardening, and exercise. Because the normative male models of human development privileged by our culture (and perhaps psychoanalysis) valorize doing over being and accomplishment over what Masud Khan (1977) termed "lying fallow," this may be quite difficult for some therapists to even consider.

The therapist's involvement in spiritual practice can result in a reevaluation of his or her goals, values, and personal expectations. This can help him or her avoid or get off what John Lennon termed the "merry-go-round" of work and pressure deriving from the societal values of success and self-worth that many therapists have internalized. What Freudenberger (1983) terms the "corrosive value influences" of capitalism (p. 84), particularly its emphasis on producing, acquiring, and consuming, are of crucial relevance in this context. With Freudenberger and Robbins (1979), I believe that "rarely can the professional immunize himself against the disease of greed" (p. 289). Defining self-worth in terms of the size of one's practice, one's material assets, or one's status in the field, for example, can promote an unconscious addiction to an overproductive and acquisitive mentality that fosters burnout. In decreasing self-centeredness and attuning analysts to the larger lifeworld in which they are embedded, spirituality can challenge these taken-for-granted values.

The prospects for lessening the unique demands the practice of psychoanalysis or analytically oriented psychotherapies places on the therapist are more favorable than we might imagine if we draw on the neglected resources of spirituality for our mental hygiene. Practicing psychoanalysis may generate deprivation and depletion, but it can also facilitate understanding and learning, compassion and connectedness, and exhilaration and joy if we have a spiritual perspective. Burnout need not be an inevitable consequence of the practice of psychoanalysis if we are willing to explore and work with the unique demands that we confront on a daily basis as practitioners. Spiritual practice can aid us in doing this. Our work and our lives will then be revivified and enriched.

REFERENCES

Cooper, A. (1986). Some limitations on therapeutic effectiveness: The "burnout syndrome" in psychoanalysts. *Psychoanalytic Quarterly, 55*(4), 576–598.

Deikman, A. (1982). *The observing self: Mysticism and psychotherapy.* Boston: Beacon Press.

Freud, S. (1937). Analysis terminable and interminable. In J. Strachey (Ed. & Trans.), *The standard edition of the complete psychological works of Sigmund Freud,* Vol. 23 (pp. 216–253). London: Hogarth Press.

Freudenberger, H. (1983). Hazards of psychotherapeutic practice. *Psychotherapy in Private Practice, 1*(1), 83–89.

Freudenberger, H., & Robbins, A. (1979). The hazards of being a psychoanalyst. *Psychoanalytic Review, 66*(2), 275–296.

Goldberg, A. (1990). *The prisonhouse of psychoanalysis.* Hillsdale, NJ: The Analytic Press.

Greenson, R. (1966). That "impossible" profession. In R. Greenson (Ed.), *Explorations in psychoanalysis* (pp. 269– 287). New York: International Universities Press.

Khan, M. (1977). On lying fallow: An aspect of leisure. *International Journal of Psychoanalytic Psychotherapy, 6,* 397–402.

Kohut, H. (1977). *The restoration of the self.* New York: International Universities Press.

Kohut, H. (1984). *How does analysis cure?* Chicago: University of Chicago Press.

Rubin, J.B. (Submitted). Psychoanalysts and their Gods. In J.B. Rubin, *The blindness of the seeing I: Perils and possibilities in psychoanalysis.* New York: New York University Press.

Rubin, J.B. (1993). Psychoanalysis and Buddhism: Toward an integration. In G. Stricker & J. Gold (Eds.), *Comprehensive textbook of psychotherapy integration* (pp. 249–266). New York: Plenum Press.

9

Psychoanalysis and Buddhism
Toward an Integration

Psychoanalysis and Buddhism each offer a map of the route to self-transformation. Psychoanalysis terms it the *therapeutic action* of psychoanalysis and in Buddhism it is called the *eightfold path* and the *factors of Enlightenment*. Both psychoanalysis and Buddhism follow a similar process that includes (1) presenting the problem of human suffering, (2) diagnosing its causes, and (3) offering a remedy. In this chapter, I shall explore the ways the Buddhist and psychoanalytic paths to transformation are synergistic, by which I mean that combining certain aspects of each could make them both more effective than if pursued alone. Since each tradition's remedy has elements that could be usefully applied by the other, studying them together could be mutually enriching. My recommendations are offered in the spirit of provisional suggestions rather than conclusive pronouncements.

The Seven Factors of Enlightenment, a classical Buddhist account of the seven qualities that comprise the Enlightened mind and the path leading to liberation, provides a model for integrating the psychoanalytic and Buddhist paths of transformation. The seven factors are (1) mindfulness, or awareness without judgment, attachment or aversion to what is happening in the present moment; (2) energy, that is the effort to be attentive and awake and to see clearly; (3) investigation, or actively probing, exploring, and analyzing the nature of things and the various dimensions of experience; (4) rapture, or the curiosity about and delight in each moment of experience; (5) tranquillity or calm, that is, "quietness of mind" (Kornfield, 1977, p. 17), "an inner kind of silence, a silent investigation rather than thought-filled" (Kornfield 1993b, p.57); (6) concentration, a state in which the mind is still, focused, and deeply immersed with laserlike "one-pointedness" on whatever it experiences;

and (7) equanimity, or a balance and evenness of mind in which one is receptive and impartial toward whatever is occurring.

There are two possible ways the factors of Enlightenment could serve as a template to integrate the psychoanalytic and Buddhist path. Self-transformation and human liberation, in the view of the Buddhist model, involves the balanced development of two sets of mutually enriching qualities: receptive, tranquilizing, and stabilizing energies and active, arousing, and energizing factors (Goldstein, 1976; Kornfield, 1979). Psychological and spiritual paths can be considered in terms of which of these factors are developed and which are neglected (Kornfield, 1979). Traditions such as psychoanalysis cultivate active qualities such as investigation (of mind and conduct), while others such as Buddhism strengthen tranquilizing qualities such as concentration and equanimity.

In Western psychology in general (with the exception of gestalt therapy) and psychoanalysis in particular, there has been a relative neglect of the value of the tranquilizing factors such as concentration and equanimity. Gestalt therapy also emphasizes moment-to-moment awareness, but because it does not cultivate the mental factor of concentration, deeply quieting and focusing the mind, its awareness does not have the penetrating depth that meditation cultivates. "Without cultivating concentration and tranquillity," as Kornfield (1993) notes, "the mind's power is limited and the range of understanding that is available is rather small in scope" (p. 58).

Buddhism is enormously effective in developing "receptive" factors such as concentration, calm, tranquility, and equanimity; it provides, as I suggested in Chapter 6, exemplary tools for focusing the mind and facilitating listening, but it has tended to neglect the cultivation of active factors such as investigation. It lacks, for example, a theory of childhood development, a developmental psychology, a developmental view of psychopathology, an explication of self-pathology with structural deficits in the self, and so forth.

There is the danger in Buddhist practice of cultivating blissful states of quietness of mind and rapture without actively investigating all aspects of one's experience such as hidden evasions of subjectivity and the cost of idealizing spiritual teachers. Altered states of consciousness do not necessarily or inevitably lead to irreversible, long-term trait changes (Kornfield, 1993a), although they sometimes do. With its capacity to illuminate and work through obstacles to spiritual practice and hidden, self-debilitating evasions of subjectivity, psychoanalysis is tremendously effective in cultivating "active" factors of mind.

The balanced development of both active and passive factors is necessary to promote human liberation. One needs to concentrate the mind and then investigate from the state of heightened clarity and inner quietude engendered by meditation.

The second way that these two wisdom traditions could be integrated will elaborate on this. In order to do this, I must provide a gloss on mindfulness, the first and most important "factor of Enlightenment." Most humans are, as the Russian mystic Gurdjieff aptly noted, "asleep"— unaware of the actual texture of their experience. They are like the president of a pluralistic democracy: they may have a general sense of what is going on, but they lack an understanding of certain crucial specifics that shape daily lived life. All authentic, noncultist psychological and spiritual systems attempt to cultivate wakefulness: attentiveness and awareness of the actual texture of experience.

In the *Satipatthana Sutra* (Thera, 1962), perhaps the locus classicus for Buddhist views on training mindfulness, Buddha asserted that mindfulness—awareness without judgment, attachment, or aversion to what is happening in the present moment—was the most important factor in diminishing unwholesome states of mind and cultivating wholesome ones. The precondition of freedom, according to Buddha, is the development of mindfulness in four areas: (1) bodily phenomena, e.g., physical sensations, breathing, body postures such as sitting, standing, lying, and walking; (2) "feelings," which refer not to what Westerners ordinarily mean by "emotions" but rather to the reactions of "pleasantness," "unpleasantness," and "neutrality" that accompany every moment of consciousness, every experience of seeing, hearing, smelling, tasting, touching, and thinking; (3) consciousness or mental phenomena, e.g., all mental states, such as anger, love, joy, lust, fear, compassion, etc.; and (4) "dharma," the universal laws underlying life, e.g., the four Noble Truths, the Eightfold path, the seven factors of Enlightenment, and the five hindrances to self-investigation.

Since all aspects of human experience are arguably contained within these four dimensions of mindfulness, broadly construed, psychoanalysis and Buddhism can be viewed in terms of which of these four areas are cultivated and which are neglected by each tradition. Psychoanalysis, for example, attends to mental phenomena to the relative neglect of "feeling," while Buddhism highlights "feeling" to the relative neglect of certain aspects of consciousness, such as emotions related to the dynamics of interpersonal relationships and intimacy. Despite the fact that meditation is not simply a solitary activity and is taught within a *relationship* to a spiritual teacher (cf. Finn, 1992, p. 168), themes related to relationships, intimacy, and sexuality are often sup-pressed or dealt with abstractly in Buddhism. This may be less true of certain strains of Buddhism such as Tantra, where the passions and sexuality are viewed as vehicles on the path to liberation rather than inevitable obstacles. These topics are, however, often not addressed directly by meditation teachers in their relationship with students. Clear guidelines about how to deal with passionate emotions, for exam-ple, are rarely offered except to counsel acceptance and detachment.

After several days of an intensive Buddhist meditation retreat, I asked one of the Buddhist teachers how to handle the wealth of insights bubbling up during meditation practice. "Don't do anything," the teacher counseled, "just let go." On one level this was sound advice because my attachment or aversion to these feelings and experiences clearly was a source of self-blindness and suffering. But given the plethora of difficulties Buddhist practitioners and teachers have experi-enced regarding various aspects of what Buddhism terms consciousness or mental life such as power, money, sexuality, and addictions (cf. Kornfield, 1993a), one wonders if the strategy of just "letting go" is a necessary but insufficient approach for the vast majority of Buddhists. The recurrent difficulties in Buddhist communities around the above issues are feedback that existing solutions in Buddhism are incomplete and that alternative strategies for addressing human consciousness and interpersonal relations might be useful.

The scandals in Buddhist communities provide a fertile context to examine these issues. The absence of data about this topic makes such an investigation difficult. A vignette about a Buddhist teacher may shed some light on how psychoanalysis might enlighten Buddhism in this area. Since my purpose is not to malign individuals, specific names are not necessary and will not be supplied.

A long-term student of Tibetan Buddhism sat next to the dharma heir of a renowned Tibetan Buddhist teacher in an airport bar watching a television talk show about AIDS. This man had been this student's teacher. He was second in command of the largest branch of Tibetan Buddhism in the United States (cf. Butler, 1990, p. 14). A guest on the talk show argued that dolls should have genital organs so that children could learn about sex and take precautions against AIDS. A commercial of a young haggard-looking woman who said that she got AIDS from her bisexual husband who was having affairs with men interrupted this program. " 'This program is trivial,' the Buddhist teacher said. 'It's cheap.' " The student thought to himself that warning people about AIDS embodied Buddhist ideals about compassionate action. He had the urge to ask the teacher why he thought the program was trivial and cheap, but pretended instead to agree with him. After all, he was the teacher and the student was the disciple. The student thought that if anybody had misperceived the situation it was probably himself. Fearing that he might sound stupid and wanting to be accepted, he was intimidated into deference. It will probably come as no surprise that he remained silent.

On the next commercial a young woman held up a hypodermic needle and said, "'I got AIDS from using this.'" The student said, "In the next commercial we'll see a guy holding up a dildo: 'And I got AIDS from using this.'" The Buddhist teacher laughed. The teacher smiled and waved goodbye as he left to catch his plane. The student's question was evaded through the joke. The student later had another opportunity to question his teacher. The teacher was leading a program for advanced students of Tantra, in which he said that if one keeps their commitment to the guru, "'you cannot make a mistake.'" The student thought to himself, "'You are going to get into trouble believing that ... that's hubris. All the people in the written history of the world who believed they could not make a mistake sooner or later got into trouble.'" The student remained silent.

A year later the student learned, with many other Buddhists, that his teacher had AIDS, kept it secret, and did not inform his sexual partners or take precautions. He infected at least one of his sexual partners. The teacher knew that he had AIDS while he was sitting with his student in the airport watching the television program on AIDS (Butterfield, 1994, pp. 3–6).

There are various ways that one might think about this actual event. It could be viewed as an illustration of what Freud and 100 years of psychoanalysis have taught us, namely, the mind's endless capacity for self-justification, self-deception, and self-blindness. Or, it could be viewed as a cautionary tale about the dangers of authoritarianism and submissiveness, and the importance of questioning "received truths" and of not making someone else, whether a Buddhist teacher or a psychoanalyst, the final authority for one's life. I shall use this experience as a point of reference to explore how psychoanalysis might enrich Buddhist understandings of the area of mindfulness that Buddha termed "mental content" or "consciousness" and the factor of Enlightenment he termed "investigation." More specifically, with its understanding of transference and countertransference, psychoanalysis could enrich Buddhist understandings of the nature of relationships in general and the dynamics of the spiritual teacher–student relationship in particular.

Many spiritual seekers in recent years have been painfully affected by the sexual, financial, and ethical improprieties of their lamas, Zen roshis, rabbis, and Christian priests and nuns. The deidealization of spiritual teachers is extraordinarily difficult to confront, since it often triggers defensiveness and anger, fear and guilt. It is usually denied, avoided, justified, or treated with condemnation and the absence of empathy. The unscrupulous behavior and indiscretions of specific spiritual teachers do not invalidate all spiritual teachings or teachers, although it cries out for attention, and thus needs to be addressed.

There seem to be four main areas of difficulty for spiritual teachers. The first involves the misuse of power. This can take a variety of forms. Spiritual teachers have decreed marriages, divorces, and lifestyles and even sometimes have abused students who did not follow their dictatorial ways.

Money is the second area of difficulty. Spiritual practice can bring great transformation and inner contentment. The gratitude many students feel sometimes leads them to give substantial amounts of money to spiritual communities. Teachers who have led simple, nonmaterialistic lives can be tempted by sudden abundance. In extreme cases, Eastern and Western teachings have been used to "make large profits, accompanied by secret bank accounts, high living, and fraudulent use of student money" (Kornfield, 1993a, p. 257).

Sexuality is the third major area of difficulty for some spiritual teachers. Some teachers have engaged in sexual "exploitation, adultery and abuse" (Kornfield, 1993a, p. 257) that has endangered the physical and emotional health of students. AIDS was transmitted to at least one student of a spiritual teacher. In 1983, an American abbot, successor to a famous Zen master, "resigned under pressure after affairs with women students, including his best friend's wife" (Butler, 1990, p. 21). In 1987, it emerged that a widely respected and supposedly celibate Korean Zen teacher "had secret long-term sexual relationships with two women students" (Butler, 1990, p. 21).

Addiction to alcohol and drugs is the fourth potential area of difficulty for spiritual teachers. A married Japanese abbot, a prominent, widely esteemed master of Zen Buddhism, suffered from chronic alcoholism, entered an alcoholism treatment program, and "apologized to his students for affairs with several women students, including a teenage girl" (Butler, 1990, p. 21). Certain Buddhist and Hindu groups have started Alcoholics Anonymous groups to deal with addictive problems.

Families have been disrupted and individual lives deeply shattered and undermined in the wake of teachers who have abused power, money, and sexuality and have engaged in addictive behavior. Exploited students demonstrate some of the same dynamics as children who are

sexually abused: shame, self-distrust and invalidation, buried deprivation and rage, guilt, fear, self-inhibition, and relational constriction.

Many of the same dynamics of dysfunctional families are discernible in spiritual communities, including denial of exploitative behavior
and the scapegoating and psychological invalidation of the victimized.
It is thus no accident that Buddhist scandals have been reacted to with
denial, defensiveness, recriminations, and blame. When Rick Fields, an
editor and author of a history of Buddhism in America, prepared an
article for *Vajradhatu Sun*, a Buddhist journal, describing the crisis in
the Tibetan Buddhist community caused by the teacher in the airport
vignette, he was forbidden by leaders of the Tibetan Buddhist community from printing it. The teacher in the vignette fired Fields when he
attempted a second time to print the article (cf. Butler, 1990, p. 18). The
censorship and suppression of information and public discussion resembles totalitarian political regimes as well as dysfunctional systems
and families.

The scandals are explained by apologists, as I suggested in Chapter
4, as an artifact of the lack of evolution of a particular teacher or as
"crazy wisdom," e.g., unconventional methods utilized by eccentric
Tibetan yogis to awaken their students, that the unenlightened cannot
fathom. Critics have explained these incidents in terms of cultural and
structural features. In terms of the former, Asian monasteries have a
strict and clear moral code involving abstinence and chastity. These
ethical guidelines and precepts have traditionally been central to spiritual communities. They legislate behavior, lessen moral ambiguity and
choice, and reduce temptation for students and teachers. Even when
spiritual communities relax ethical strictures such as allowable drink-
ing in China or Japan, there are "strict cultural norms for the behavior of
the teachers" (Kornfield, 1993a, p. 259). Members of the community
support this by, for example, dressing modestly so that the teacher is
shielded from sexual stimulation. These cultural norms are often an-
athema to and absent from late twentieth-century American culture
where the freedom and fulfillment of the individual, rather than abstract
ethical guidelines, are usually viewed as sacred. Spiritual teachers may
commit improprieties, according to this sociological perspective, when
they confront a cultural context in the United States without institutional support for a spiritual way of life. This begs the question of why a
spiritual teacher would act immorally toward a student even when
external ethical guidelines were not in place. It also seems to remove
individual responsibility for ethics, as if one acted immorally because
there were no guidelines. This perspective offers a twist on the child or

adolescent who tells his or her parents or therapist that "my friends made me do it": "the lack of community-wide ethical guidelines made me do it."

Power, money, sexuality, and addictions are not included in the training of most spiritual teachers. In fact, an exploration of sexuality, power, and money is explicitly excluded from most spiritual systems and practices (cf. Kornfield, 1993a, p. 258). This can result in a "compartmentalization" within the teachers, by which I mean, teachers may be skilled at teaching meditation, conducting Zen koan practice, or guiding students in loving kindness or visualization practices, yet be uncomfortable with their own feelings, body, interpersonal relationships, or intimacy. In a study of 53 Zen masters, lamas, swamis, and/or their senior students about their sex lives and the sexual relations of their teachers Kornfield (1993a) found, among other things, that "there were many more teachers who were no more enlightened or conscious about their sexuality than everyone else around them. For the most part the 'enlightenment' of many of these teachers did not touch their sexuality" (p. 259).

The transplantation of Buddhism to the West places Buddhist teachers who grew up in hierarchically stratified cultures and were culturally shielded from certain facets of life in the West, such as sexuality and money, and are conditioned to accept set and inflexible sex roles in a psychologically unsettling position when they confront students in American culture characterized by more egalitarian relations (Roland, 1988), fluid roles, and less formal modes of relating. Buddhist teachers in America may thus be exposed to sexual and financial temptations that may not have existed in their countries of birth.

The guru–disciple relationship is often authoritarian. Obedience and surrender are the prime virtues (Kramer & Alstad, 1993, p. 15), and those in power (the spiritual teacher) usually demand submission and agreement and those in the more deferential role (the student) assume that they must unquestionably obey the one in power. Both control and surrender have deep attractions or what Freud termed secondary gains, as well as dangers. Submission provides direction, meaning, and security for the spiritual seeker, while the power and adulation the teacher commands may feed his or her pride and self-esteem. The mutual attractions can create "disguised collusions" (Kramer & Alstad, 1993, p. xiii) among both participants in which they are both invested in the perpetuation of the existing relationship.

The potential for mental control, submission, and coercion has the potential to breed corruption. When the teacher is essentially unchallengeable, as she or he usually is in most spiritual circles, the student is in an essentially submissive position and the teacher is immunized to feedback. The presence of a relationship structured by dominance and submission and the absence of feedback about how it operates and what is not working insures that self-deception, self-inflation, and corruption are quite possible and that change and transformation are unlikely. In addition, since the attractions of surrender and dominance are rarely, if ever, examined in spiritual circles, their perpetuation and continuation seem inevitable.

Let us return to the case of the Tibetan teacher. From a Vajrayana or Tibetan Buddhist perspective, "passion, aggression, and ignorance, the sources of human suffering, are also the wellsprings of enlightenment. Afflictions like AIDS are not merely disasters but ... opportunities to wake up" (Butterfield, 1994, p. 7). With its emphasis on nonjudgmental attentiveness, Buddhist practices seem to offer the possibility of examining scandals and difficulties in either an individual's practice or a spiritual community in an enlightening manner.

What would a Buddhist interpretation of the teacher's behavior be? From within a Buddhist perspective, the teacher needs to wake up to the web of desire, attachment, craving, and ignorance he is ensnared in. He is attached to sense desire and perhaps afflicted by greed and delusion. Desire is conditioned by ignorance. He takes what is impermanent, insubstantial, and unsatisfactory as permanent, substantial, and satisfying. He needs to renounce the desires he is driven by.

Morality (*sila*) is one of the three pillars of the eightfold path, the Buddhist map for facilitating change. Certain facets of the eightfold path that are explicitly concerned with ethics such as right speech (avoiding lying, gossip, and harmful, derogatory speech) and right livelihood (engaging in work that helps rather than detracts from life) provide an opportunity for what might be termed a "life-wide" praxis. By this I mean that the "stage" of working on oneself in Buddhism, to borrow from Shakespeare, is all the world of the person's life ranging from how one speaks to others to the kind of work one does. Nothing in psychoanalysis makes it adverse to investigating any aspect of human psychological life. But certain aspects of speech and work, for example, the pettiness, backbiting, and lack of civility permeating analytic institutions, politics, and professional meetings and conferences, often seem to escape analytic scrutiny. In suggesting that the therapeutic stage is all the world, Buddhism is pinpointing the value of examining the totality of one's life and potentially enlarging the scope of analysis. One's whole life, not merely specific hours during the week, might potentially be part of one's practice; grist, we might say, for the meditative or therapeutic mill. But because of the neglect of both the factor of enlightenment

termed "investigation" and the area of mindfulness termed "consciousness" or "mental content," Buddhism is also inhibited in optimally
investigating the totality of a human life.

Buddhist explanations of pathology, for example, unconscious attachments based on what classical Buddhism terms the "three poisons"
(greed, hatred and delusion) (cf. Goleman, 1977), leave us with several
unanswered questions about the Tibetan teacher. If desire is associated
with the feeling that one wants to have or to possess, why does one—the
Buddhist teacher—want to possess someone or something? Why *did* the
Tibetan teacher feel desire, greed, and delusion? What is the unconscious meaning and purpose of craving or being deluded? What are the
secondary gains of such behavior? What is it in the service of? Are there
earlier, troublesome relational configurations that he is recreating? What
gets warded off or avoided when he acts this way? Who in his family of
origin might get protected? Why do humans, including this teacher,
cling to self-damaging ways of being?

Since there might be more to pathology than "attachment" to illusory views of subjectivity and greed, hatred, and delusion, adopting the
Buddhist view alone could potentially eclipse these and other important questions about the nature and cause of pathology. Let us take a
brief detour into psychoanalytic understandings of transference and
countertransference in order to illustrate how psychoanalysis' skill in
investigating mental content might be enriching to Buddhism.

Many spiritual teachers may assume, as many analysts have, that
they stand outside the student and observe and relate to him or her
objectively. Many twentieth-century scientists, philosophers, historians, psychoanalysts, feminists, anthropologists, and literary theorists
have taught us that objectivity is an illusion. There is no immaculate
perception or theorizing. Observation is always riddled with subjectivity, including the personal history, theoretical predilections, wishes,
and needs of the observer. Buddhist teachers, like analysts, are thus
more like what Harry Stack Sullivan (1953) termed participant–observers
than neutral scientists.

Since Freud, psychotherapists and psychoanalysts have recognized
that patients' recurrent ways of approaching self and others usually
manifest in analysis and in the analytic relationship in the form of
transference. Transference has been defined in a variety of ways since
Freud used the term to describe the way one of his patients experienced
him as if he were a "new edition" of a formative earlier figure in her life.
The feelings, thoughts, and fantasies attributed to Freud were out of
proportion to the present context and were in fact related to important

figures from the patient's personal past. For our purposes here, I will define transference—with Gill (1982), Levinson (1983), Wachtel (1977), and Stricker and Gold (1993), among others—as the way one's interactions in the present are shaped by one's inherited and internalized past, enduring patterns of relatedness, and characteristics of the present relationship. In transference the analyst's world and the therapeutic relationship are assimilated into the patient's worldview.

Transference is not simply the illusory apperception of another person—what the patient *attributes* to the analyst, whether the expectation of criticism, protection, deprivation, or seduction—but what the former unconsciously attempts to *enact* with the latter. The spiritual teacher for example, will not only be experienced as if he or she were the student's parent or parental surrogate (whatever that might mean for the student), the teacher will unconsciously be induced to behave as if he or she is the student's parent. The relationship between the spiritual teacher and the student thus becomes the stage for the student's hopes and fears, dreams and dreads. Characteristic wishes, modes of self-protection, and patterns of relatedness may be enacted in such relationships.

The transference, for Freud, was both an obstacle to and a vehicle for the treatment. When transference remained hidden or unanalyzed, it could and did disrupt the treatment. Studying transference affords the opportunity to examine one's characteristic ways of thinking and relating. When it is analytically illuminated and worked through, it is a powerful force in freeing the patient from the grip of his or her past as well as opening up possibilities for new modes of relating and living in the present.

With the relational turn in psychoanalysis, the greater concern with the impact of relationships on human development, and the process of change in the therapeutic process, an increasing number of analysts have utilized their own experience, including countertransference as a tool for understanding a patient's transferences and illuminating his or her subjective reality. Countertransference, like transference, is not one unitary thing. Freud recognized that it created blind spots in the analyst and interfered with the treatment. One way that it has been viewed is the analyst's reaction to the patient's transference that interferes with the treatment. Countertransference could be viewed as the way the analyst assimilates the patient's experience and the analytic relationship into the analyst's categories.

Freud never took the step with countertransference that he took with transference of recognizing it as a useful tool in treatment. Subsequent analysts such as Little, Racker, Heimann, Sandler, Levinson, and

Mitchell, among others, have recognized it as an indispensable ally in the treatment.

Buddhist practitioners also relate to their teachers in ways that are characteristic of their personalities, self-deficits, unconscious conflicts, patterns of relatedness with others, and modes of caring for or neglecting self. Many students come to spiritual practice, as Engler (1986), among others, has noted, with self-disorders and pathology; that is a fragility and disturbance in their sense of self, fluctuating and impaired self-esteem, a feeling of not being real, cohesive, or temporally continuous. In addition, such people may feel a sense of emptiness, enfeeblement, fragmentation, and depletion (Kohut & Wolf, 1978). People attracted to meditation may also feel as if they are bad and may have powerful unresolved yearnings for acceptance and love from childhood. With its emphasis on the doctrine of "no-self," Buddhism (as well as other Eastern contemplative disciplines) may be attractive to such students who experience themselves as feeling empty, hollow, or unreal; as having no self. Engler (1986) described a student in a college class he taught on Buddhism who illustrated some of these dynamics. The student equated his state of self-emptiness with Enlightenment. He maintained that he lived in a state of egolessness so that there was no need to meditate. He alternated between idealizing and devaluing Engler. In one class, Engler was an omniscient teacher who appreciated his genius. In the next class, he was yet another disappointing and limited figure who failed to appreciate his talents and insights.

The doctrine of "no-self" not only resonates with a student's own experience of inner emptiness, lack of self-cohesion, and the absence of a sense of temporal continuity, it can be an attempt to repair a vulnerable self through grandiose achievement. One "achieves" the summum bonum, Enlightenment, a state of perfection, invulnerability, and self-sufficiency.

Engler (1986) maintains that two major groups become interested in Buddhism and attend retreats: those "in their late adolescence and the period of transition to early adulthood, and those entering or passing through the mid-life transition" (p. 29). One attraction of Buddhism for such people may be that it gives them a moratorium on certain developmental tasks relevant to their respective stage of the life cycle.

Buddhism may also be appealing to people who fear autonomy or intimacy. Lacking a belief in one's own value and competence, one can escape the burdens of being responsible for one's life by passively surrendering to religious traditions, communities, or gurus who provide direction and guidance for how to lead one's life.

Pursuing spiritual practice can also sanction the avoidance of relationships. One avoids fears and risks associated with involvement in relationships, dangers such as rejection, abandonment, engulfment, and suffocation, by remaining "detached" and uncommitted to anyone except the spiritual teacher, community, or practice. One Buddhist monk, for example, explained to an interviewer that he avoided women because they evoked lust, which interfered with quieting his mind for meditation practice (cf. Fauteux, 1987).

Spiritual practice is sometimes used to ward off grief and mourning as well as intimacy. Through meditation practice, one can quiet the mind, detach from disturbing emotions, and thereby momentarily obviate the need for experiencing or working through disturbing feelings.

Students plagued by guilt may be drawn to spiritual practice because it can purge them of a conviction of inner badness, fuel self-depriving and self-punishing tendencies, like Steven in Chapter 5, and help them experience the punishment that they imagine they deserve, thus atoning for their imagined transgressions.

Wounded, isolated, and lonely, some spiritual practitioners seek the acceptance and love that was not forthcoming from primary caregivers. "No matter how many communes anybody invents," notes Margaret Mead, "the family always creeps back in" (quoted in Kornfield, 1993a, p. 261). Spiritual seekers hope to receive from their spiritual teacher or community what they sorely lack and may never have experienced in their family of origin.

Two characteristic transferences in psychoanalytic treatment that Kohut postulated a person utilized to bolster an endangered or depleted self and thus heal a fault line in the self—mirroring and idealizing ones—may illuminate what sometimes occurs in the teacher–student relationship in Buddhism. Students may, for example, hope the teacher will provide either confirmation and approval or be an idealized source of strength, calmness, and wisdom that the student can merge with.

The Eastern literature has anecdotes of teachers and masters interacting with students and disciples in ways that are designed to help them recognize characterological patterns. But the use of transference is unsystematic and sporadic. Buddhist teachers tend, as one teacher of Buddhism aptly notes, not to "pay much attention to transference aspects of the teacher–student relationship" (Engler, 1986, p. 37). This both shapes and renders invisible aspects of the relationship and it inhibits both teacher and student from utilizing what occurs in the relationship as a vehicle for self-knowledge. Exploring whatever occurs in the relationship, on the other hand, could be enriching for both participants.

Psychoanalysis can enrich Buddhist understandings because in the psychoanalytic situation transference (as well as countertransference) can be systematically analyzed and utilized to illuminate ways of being that may either go unnoticed or be submerged in Buddhism, such as the idealization of the teacher and the concomitant submission of the student (Deikman & Tart, 1991). In Buddhism this dynamic may remain unexamined and the student's self-devaluation and deferentiality may never get resolved and may play itself out in various other relationships. Because of the relative neglect of the area of mindfulness that Buddhism terms "mental content," Buddhism is hamstrung in exploring and illuminating teacher–student relationships or strife in Buddhist communities for at least two reasons: First, Buddhism, as I mentioned earlier in this chapter, lacks a theory of human development; it does not illuminate the impact of individual psychological history on subsequent development. Second, it also neglects the vicissitudes of illusion and blindness operative in human relationships, the terrain covered by psychoanalytic notions of transference.

Let us return to the case of the Tibetan teacher. He felt something in relationship to the student(s?) he slept with, either emanating from the past or evoked in the present. There are many possibilities. Without more knowledge, it is impossible to say whether he felt lust, rage, deprivation, revenge, or a combination of other feelings in his relationship with the student he infected with AIDS. We also do not know anything about who was interested in whom, who initiated contact, whether the interest was asymmetrical or mutual, or whether the intimacy was based on coercion, voluntary participation, or a combination of each. All we know is that the sexual contact between the Tibetan teacher and the student was literally deadly.

Lacking an awareness of transference or countertransference, the Buddhist teacher would have had great difficultly in exploring the dynamics of the relationship. Needless to say, there are many ways one could think about what happened. Here is one way a psychoanalyst might think about what happened.

The analyst, according to Heimann (1950, 1960), has to be able to draw on the feelings and reactions that are stirred up, as opposed to acting on them in order to utilize them for understanding the patient. Meditation teachers need to use their reactions and responses to their students, whether, for example, affection, aggression, or disinterest, as a tool to understand their students and themselves. When the meditation teacher, like the analyst, avoids the reductive options of either phobically pushing passionate feelings away (as some meditation teachers may do), or acting them out (like the Tibetan teacher did), then a third possibility arises; namely, that one could become curious about what was occurring between the teacher and the student and within each person. Was the Buddhist teacher trying to sexualize needs for emotional connection and validation? Was he reenacting a betrayal he had experienced earlier in his life? Was he "screwing" others the way he had been screwed? Was he denying the reality of his impact on the student he infected so that he was protected against identifying with the student who unconsciously represented a disavowed facet of himself? Did he consciously or unconsciously assume that the world was inhabited by potential abusers and that he needed to turn the tables on them before they mistreated him? Was he attempting to pass on the AIDS virus in order to create a psychological twin in order to lessen his loneliness?

Treating a student's sexual overtures or idealization of the teacher and self-devaluation or a spiritual teacher's desires as grist for the investigative mill, something to understand rather than something to act on or avoid, would make it more possible to raise questions about what was transpiring between teacher and student. It would also illuminate the student and teacher's habitual ways of caring for and relating to themselves and others. Increased self-awareness, relational attunement and sensitivity fosters a deepened spiritual practice. Psychoanalysis' skill in investigation illustrates the second way the psychoanalytic and Buddhist paths might be synergistic.

In the next section, I shall discuss a case of a Buddhist in psychoanalytic treatment. I shall focus on those aspects of the treatment that will illustrate the approach to integrating psychoanalysis and Buddhism that I am advocating.

> East is East and West is West and never the twain shall meet. (Kipling, 1942, p. 878)

> I too have ropes around my neck, I have them to this day pulling me this way and that, East and West, the nooses tightening, commanding choose, choose.... Ropes I do not choose between you.... I choose neither of you and both. Do you hear? I refuse to choose. (Rushdie, 1994, p. 211)

My work with a man I shall call Albert, an affable humanities professor in his late 20s who suffered from conflicts over individuation and success, excessive self-judgment, diminished self-esteem, inauthenticity, compliance, and a pervasive sense of directionlessness and meaninglessness, refusing to choose between either East or West, Buddhism or psychoanalysis, or rather choosing both (and seeing where the twain both did and did not meet), has made all the clinical difference. Albert was an only child who was raised as an agnostic. In the early stages of treatment he described his mother as caring and devoted. As treatment proceeded, other images of her emerged. He later saw her as a rigid person who was more concerned with everyone conforming to her view of reality, which included how her son should act and be. She was deeply committed to banishing subjective life and she demanded that everyone live in accordance with her narrow view of reality. He felt that she lived in a "fortune cookie" universe in which her "shoulds" were lionized. Cliché responses ("you must have felt badly when so and so died") replaced genuine affective engagement. Albert felt impactless and nonexistent in her presence. His mother was like a "fencer who parries everything I say." Instead of being affirmed and validated by her, he felt nonengaged and invisible.

From Albert's account, he had a distant relationship with his father, who he experienced as intelligent, detached, critical, and passive. His father submissively went along with his wife's way of living and relating. His father's blind loyalty to his wife resulted in Albert's never feeling understood or supported by him. His father could not sustain interest in him. Albert repeatedly cited numerous examples of his father changing the subject whenever Albert shared some musical or academic accomplishment. Albert gradually realized that although his father had made him feel that he was incompetent and inadequate, in actuality his father avoided many areas of life that he let his wife take responsibility for.

As Albert described the way his parents laundered communication of all subjective meaning, vaporized conflict, and nullified authentic selfhood, while appearing to exude empathy and significance, I was reminded of Bollas' (1987) suggestive account of "normotic illness" (p. 135). Such people are "fundamentally disinterested in subjective life" (p. 136). They have a "disinclination to entertain the subjective element in life whether it exists inside ... or in the Other. The introspective capacity has rarely been used ..." (p. 137). Normotics are invested in "eradicating the self of subjective life" (p. 156). "The normotic flees from dream life, subjective states of mind, imaginative living" (p. 146).

Albert's parents did not see themselves or him as subjects capable of introspecting, feeling, desiring, or playing. Rather, he was coerced into accommodating to their preexisting viewpoint on reality. To stay connected to his parents, he needed to hide his subjective life. He "harmonized" with his parent's normotic view of reality, so as not to be like "an astronaut cut off from home base in outer space," because that was the only way of keeping alive the hope of being emotionally related to them.

His parents were sorely unresponsive to his inner reality and failed to encourage his uniqueness from emerging or flourishing. In fact, they encouraged both false self compliance with their narrow mode of being by emotionally rewarding submissiveness and conventionality and discouraging authenticity and individuation.

He felt like a powerless pawn in their play designed to whitewash reality and his subjective life. Albert developed a private subjective world of depth and richness, but he had great difficulty believing in its validity and sustaining his commitment to it. The price of conforming to and accommodating his parents wishes was to bury his own sense of how he should live. He kept alive the tenuous hope of being accepted by his parents by banishing huge parts of his self through subverting and obscuring his own "voice." This led to an excessively limited view of himself and his capabilities. What he wanted lacked significance to him and he felt that his life was not his own. He experienced himself as an "ellipsis/virtuality," with certain important parts of his potentialities dormant. This left him feeling self-doubtful and directionless.

He had what sounded at first like vital and interesting male friendships characterized by openness and depth. As time went on, material emerged suggesting that several of these friends were either narcissistic or emotionally needy. His male friendships seemed organized so that he would provide a great deal of psychological sustenance. Relationships with women were often characterized by a great desire to be related, a self-nullifying attention to their emotional needs, which left out his own needs, and a stance of pseudo-incompetence. He dove into the world of scholarly pursuits and Buddhist practice, avoided the world of practical affairs, and set it up so that the women he was involved with would take care of these matters, which secretly left him feeling incompetent and unable to have the kind of partnership he desired.

His competence and expertise were also carefully hidden from others in his academic work. He assumed that his competence would either be disregarded by those he valued or threatening to less able colleagues. He was afraid that competence and self-assertion would result in his being shunned and "orphaned." He simultaneously complied with the needs of others and related in a deferential manner and secretly pursued interests and avocational pursuits, such as Buddhism, that did not fit into the mainstream of the familial, professional, and social worlds he inhabited. That his abilities were not appreciated fed his sense of invisibility. He tended to become involved in projects with colleagues who were not his equal, which reinforced his sense of deprivation.

We observed in his life what Pontalis termed a dialectic of "death work" in which "parent and child develop a reciprocal preference for maintaining an unborn self" (Bollas, 1987, p. 144). He described the process of denying his own needs, subordinating himself to what others wanted, and neglecting his own goals as "Albertlessness." It involved both deferring to the needs of others and "straitjacketing" himself. In later stages of analysis, he described this as being "buried alive."

In the beginning of treatment, Albert was affable and compliant. Initially he did not talk about Buddhism. As we explored his tendency in the transference to conform to me by attempting to speak my language, his fear that he would be alone and invisible if he did not accommodate to those around him, including me, emerged. As we understood two of the dangers he anticipated—that I would be like his critical father or his usurping mother—material about Buddhism emerged. After several months in treatment Albert reported on his recent experiences with Buddhist meditation. He had just returned from pursuing intensive meditation practice at a Buddhist retreat center and found it enlightening and inspiring. He practiced Vipassana meditation.

He described the way meditation practice cultivated heightened attentiveness and self-awareness. The heightened awareness helped him begin to be able to know his own reactions more easily and steadily. He then accommodated less readily to the needs of others.

He also emphasized the way ethics seemed more central in Buddhism than psychoanalysis. From what he had observed, the five teachers of meditation that he had studied under seemed to have a more expansive view of the world and morality than the therapists he had encountered socially and professionally. He illustrated the more panoramic view of ethics by a story about Gandhi. A man went to a talk that Gandhi gave with the goal of killing him. Moved by the power of Gandhi's teachings, he shelved his plan. After the talk he prostrated himself in front of Gandhi and told him of his change of heart. Gandhi's response to this man, a potential assassin, was "How are you going to tell your boss about your failed mission?"

My own stance toward his involvement with Buddhism, which was never stated explicitly but was probably implicit, was to approach it with neither unequivocal reverence nor defensive rejection. Heeding Freud's (1921) warning to avoid "two sources of [interpretative] error— the Scylla of underestimating the importance of the repressed unconscious, and the Charybdis of judging the normal entirely by the standards of the pathological" (p. 138), I attempted to avoid the twin dangers of a priori pathologizing, which would reject Buddhism automatically, or unconstrained idealization, which would accept it uncritically. If psychoanalysis has too often been reductionistic in pathologizing religious phenomena, then challenges to psychoanalytic reductionism within religion and nonanalytic Western psychologies have often fallen victim to a reverse pitfall: accepting religious experience too uncritically and simplistically. The complexity and depth of religious phenomena is eclipsed in both approaches and religion and psychoanalysis are then both impoverished.

The Eurocentrism pervading psychoanalysis, that is, the tendency to treat European standards and values as the center of the intellectual universe, has often blinded analysts to the potential clinical value of Buddhism as well as other spiritual traditions. In our clinical work it became evident that Buddhist meditation practice helped Albert in several ways. It helped cultivate enhanced self-observational capacities, e.g., it increased his attentiveness and self-awareness. When he discussed relationships, for example, he demonstrated great subtlety of insight as to the possible patterns of interaction and the hidden meanings and motivations that might be operative. This enabled him to do such things as track his reactions to me and others and often detect inchoate perceptions and fantasies. While meditating, for example, he became aware of formerly disavowed feelings of betrayal at the way his mother "gaslighted" or mystified him.

His self-introspective abilities seemed to increase as his involvement with meditation deepened. Since such self-awareness is also central to psychoanalysis, it seems impossible to ascertain the relative influence of each tradition in promoting it. One factor that suggested to me that meditation practice played an important role was the fact that several times Albert returned from meditation retreats the day of one of our sessions and seemed unusually attentive to nuances of his inner life, such as latent motives, formerly disavowed intentions, and so forth.

Not only did his receptivity to internal and interpersonal life increase, his attitude toward his experience changed. The meditative spirit of attending to experience without judgment or aversion gradually replaced the self-critical stance exemplified by his father.

Not only did he become significantly less self-critical; he also became less "attached" to experience, less addicted to pleasing experiences and less afraid of painful ones. Attachment, as Buddhism tirelessly emphasizes, causes suffering. Albert, like the rest of us, suffered when he was attached, when he wanted things to be his way, when he tried to control experience. As he meditated more, he seemed to become less attached. He accepted change. He let things happen. He flowed with experience, rather than demanding it be his way or prereflectively utilizing habitual coping strategies. For example, after he had meditated for some time, he reported the following incident while visiting his parents for dinner. He witnessed the same disturbing dynamics that had permeated his childhood. He experienced rage at his father's belittling treatment and his mother's deadening normotic behavior. But instead of reacting by either tuning them out or mindlessly fitting in, as he had before he meditated, he was able to experience anger and deprivation without resorting to either avoidance or self-hurtful or self-anesthetizing behavior. Detachment helped him become less caught in habitual or conventional reactions and attitudes. This resulted in decreased defensiveness and more openness to experience.

He had a highly developed capacity for tolerating and living in and through a range of affects without having to foreclose or simplify either the confusion or the complexity. He was able, for example, to examine such things as ambivalence, anger, and perplexity without judging himself, reducing the complexity of these experiences, or clamoring after premature understanding. As he accepted himself more, emotional warts and all, he developed deeper acceptance of others.

He then experienced a fluidity, an openness of being very different from the normotic rigidity of his parents that foreclosed experience. His lightness of being was not simply born of accommodation and submission, although parts of both may have been operative. As a result of all this, his suffering decreased, both the "neurotic misery" that Freud felt psychoanalysis addressed *and* the "common human unhappiness" he claimed psychoanalysis could not resolve.

Buddhism's dereified conception of subjectivity was also liberating for Albert. By continually revealing to him on an experiential level the fluidity and self-transformative facets of self-experience, the way consciousness changes moment after moment, the practice of meditation offered him a more fluid way of experiencing subjectivity. Meditation taught him that the apparently unified image that he and his parents constructed about him was, in part, illusory. This homogeneous image of the self hid its complexity. An unreified and unconstricted sense of subjectivity was thus facilitated. By offering a less reified view of self Buddhism helped Albert challenge the deadening clutches of his parents and his own preset, one-dimensional, normotic view of himself.

My own knowledge of Buddhism's view of self-experience alerted me to the dangers of self-reification, which helped me recognize and avoid the subtle self-commodification that occurred in his family, his relationship to himself, and periodically in his previous analysis and our analysis. In these four situations he was made into a thing; a means rather than an end; an It rather than a Thou. Self-commodification occurred in his family when he was narcissistically viewed as an extension of his parents. His parents were more concerned with making him into the sort of person they needed him to be for them to feel better about themselves than to facilitate his own differentiated development.

In our analysis a pattern emerged that had apparently shadowed his previous analysis and probably our work in which he would almost instantaneously bury negative feelings, e.g., the feeling that I had not fully appreciated a particular accomplishment, and shift the analytic focus onto understanding what he had done wrong. His self-bashing was sometimes quite subtle and not so easy to detect both because it was so automatic and because the shift in focus he initiated had some plausibility; in other words, he would direct us toward areas that seemed to be troublesome for him. For example, after one session in which he had discussed what he had learned from both Buddhism and psychoanalysis, he suggested that we examine the way in which he sometimes undermines himself and hides his intellectual potential. He suggested that his convoluted presentation of his ideas during the previous session could provide a clue to what he needed help with in this area. He suggested that his presentation was symptomatic of a difficulty he sometimes had in his intellectual work. My experience of the previous session was that he had made some subtle and important points about both psychoanalysis and Buddhism but that some of his ideas were still inchoate. I also felt that I had not offered him much assistance in further clarifying the material. After I commented in the current session about one of his ideas that I believed was quite interesting, he realized through the contrast between my more zestful response in the current session (compared to the previous session) that the previous session had been disappointing for him. Further exploration revealed that he was disappointed that I had not been more effusive about his ideas. He realized that when he suggested at the beginning of the current session that we examine the negative way he thinks that he was both obscuring his disappointment in me for not being more enthusiastic about his intellectual work and making himself into a "commodity" to be worked on and improved. He indicated that he felt that this kind of process pervaded his previous analysis. The Buddhist dereified view of subjectivity, particularly its appreciation of the fluidity of consciousness and the importance of the transient, episodic mental states arising moment after moment, helped both of us to be sensitive to the way self-commodification occurred in our work and in his life outside analysis, which led to a more unconstricted view of himself.

Non-self-centric moments of consciousness and experiences of self-transcendence, in which there is heightened attentiveness, focus, and clarity, attunement to the other as well as the self, non-self-preoccupied exercise of agency, a sense of unity and timelessness, and non-self-annulling immersion in whatever one is doing in the present, became more apparent as attention was paid to the fluidity of subjectivity.

Because psychoanalysis lacks a psychology of transcendent states these modes of being do not appear on its map of human development and are assumed to be pathological. The ways they might represent expanded states of consciousness qualitatively different from more archaic states of nondifferentiation and enhance self-experience are thereby eclipsed.

With its more expansive, less self-centered view of self, Buddhism suggested possibilities of selfhood and morality that went beyond the narrower views of either psychoanalysis or his family. That psychoanalysis, even relational versions, treats the Other reductively is suggested by its notion of the Other as an object, its failure to articulate a workable theory of intimacy, and its egocentric conception of responsibility. It may not be accidental that psychoanalysis calls the Other an "object" and relations with others "object relations." The word *object* connotes a thing and predisposes one to adopt a depersonalized view of (cf. Buccino, 1993, p. 130) and narcissistic relation to others, in which one focuses on what the other needs to do for the self rather than what the self might do for the other. The Other tends to be viewed, in psychoanalysis, as an object, not a subject.

The narcissistic view of relationships emerges in the psychoanalytic vocabulary of bad objects, part-objects, need-satisfying objects, and self-objects. It is not surprising that psychoanalysis lacks a nuanced and compelling account of emotional intimacy among egalitarian subjects. For how could there be an adequate account of intimacy when the other is seen mostly in terms of what it does (or does not do) for the self? In a theoretical universe in which "relationships are secondary phenomena and emotions are derivative, love will never be discovered" (Gaylin, 1988, p. 44). The complexity of intimacy and love cannot be adequately explained when there is a valorization of humans as self-centered and hedonic monads accountable only to the dictates of their own personal tastes and dispositions.

From Thai Buddhists irritated about religious demands of Muslims in Thailand to Californians who resent the influx of immigrants, the claims and strains of Otherness are a central facet of the world. Understanding and getting along with the Other, whether in or outside of analysis, becomes more crucial in a world of shrinking/shifting borders and the globalization of communication in which we all confront increasing scarcity, selfishness, intolerance, disconnection, alienation, and nihilism.

The psychoanalytic neglect of Otherness affects its view of morality as well as its conception of interpersonal relationships. In this section, I shall attempt to demonstrate that psychoanalysis both compromises our understanding of morality and complicates and enriches spiritual practice.

Psychoanalysis may now be very far from Heinz Hartmann's (1960) assertion in *Psychoanalysis and Moral Values* that it is a clinical procedure without moral considerations. While Hartmann's position seemed to echo Freud's (1933) claim that psychoanalysis does not posit any values beyond those within science as a whole, moral questions pervade it, from value lacunas of patients seeking our help (cf. Meissner in Panel, 1984) to the implicit and explicit values in our theories and practices. From the metapsychologies we utilize to the decor of our offices, values impinge on psychoanalytic theories and clinical work. Once it is recognized that the moral neutrality extolled by Hartmann is neither possible nor desirable, the question becomes: "What values are operative in psychoanalysis?"

There seems to be an ambivalence within psychoanalysis concerning values and morality. Let me cite two examples. Conscience, ideals, and values—the terrain covered by the concept of superego—is one of three pillars of self-experience for many analysts since Freud. The rigidity and harshness of the superego is presumably lessened in the course of successful treatment. Value neutrality, on the other hand, is still deeply valued by many analysts who believe in the possibility of and continue to aspire to being scientific. Yet, to know what a cure is, to know what recovery would look like, the analyst must already have a vision of what the good life is (Phillips, 1994). Such a vision places us squarely in the land of morality. Second, psychoanalysis does not "espouse a single ethical point of view" (cf. Gedo, 1986, p. 214) and lacks normative criteria for comparing competing systems of morality, yet it can bring more reason to bear on one's asocial, self-centered drivenness, lessen irrational conscience, and improve our standard of conduct (cf. Gedo, 1986, p. 207).

Because ethics and tolerance also involve decentering from the centrality of our own cherished viewpoints and taking into account those of others (cf. Varela, 1984), psychoanalysis offers a necessary although insufficient perspective for thinking about morality. For the greater self-acceptance and lessened narcissism that often develop as a result of analytic treatment exist without the larger perspective fostered by spiritual perspectives, wherein one's own experience is viewed as a *part* of a more encompassing reality. A psychoanalytic view of morality based on an individualistic sense of self leads to a morality rooted in the neglect of the other, which does not provide an adequate framework for ethics.

Moralities rooted in religious thinking, whether, for example, in the form of the ethic of Christian fellowship and love or Buddhist egoless compassion, may encourage less self-centered views of self and more humane conduct. Freud complicates the moral landscape and the vocabulary of moral reflection at one's disposal with his monumental "decentering" of self-mastery resulting from psychoanalysis' demonstration that the ego "is not master in its own house" (Freud, 1917, p. 143). The availability of a richer vocabulary of moral distinctions fosters greater sophistication and sensitivity in our moral deliberations. It is more difficult, if not more self-deceptive, to take religious claims about doing good and acting benevolently at face value in a world in which generosity may hide self-aggrandizement and self-deprivation. Nonpsychoanalytic spiritual conceptions of self and morality tend to neglect the psychological complexity of selfhood and ethics, for example, the multiplicity of conflicting motives that may underlie a particular action.

The word *responsibility* hardly appears in psychoanalytic discourse. Despite the relational turn in psychoanalysis in the last 15 to 20 years, the morality underpinning psychoanalysis, even a relational one, is based on a one-person, nonrelational model of human beings. While contemporary psychoanalysis charts relational influences on development and treatment, it discourages a relational perspective on moral responsibilities. In contemplating action/moral decisions, we unconsciously import a one-person perspective on morality, by which I mean when patients struggle with ambivalence about a relationship or moral dilemma—should I, for example, remain with or leave my partner? Should I allow my aging parent to move in?—we psychoanalysts usually do not ask "What is right?" but "What do *you* feel, want, or need?" Our very question about what the lone individual wants or needs predisposes us to think of single individuals, not of people inextricably involved in a network of relations.

In asking not what the other can do for the self, the question his parents and psychoanalysis focused on, but rather what the self might do for the other, Buddhism offered Albert an alternative vision to the solipsistic, normotic world of his parents. Albert tended to view morality in a more complex and nuanced way because of his concern with the other as well as the self.

Psychoanalysis needs Buddhism in order to become less excessively narrow, less self-centered, less focused on the needs and wishes of the separate self. Buddhism teaches us that when we think about action, choice, moral commitment, and relational affiliations, we need to focus on the other as well as the self.

Narcissism, Buddhism has taught us, is at the root of both human suffering and evil. It has recently been theoretically claimed, although never clinically demonstrated, that "meditation is a means of indefatigably exposing this narcissism, of highlighting every permutation of the self-experience so that no aspect remains available for narcissistic recruitment" (Epstein, 1995, p. 134). The scandals in Buddhist communities involving Buddhist teachers (Boucher, 1988), those individuals presumably selected by enlightened Buddhist masters to teach in part because they represent the highest level of spiritual realization and ethicality among Buddhist practitioners, raise telling questions that Buddhists and Buddhistically sympathetic Western mental health professionals have rarely, if ever, addressed or illuminated. Why, for example, is there so much unconsciousness, egocentricity, and amorality in a tradition that prizes and focuses on self-awareness, selflessness, ethicality, and compassion? Why do years of meditation practice not prevent this sort of exploitative conduct? Is narcissism really eradicable?

If the Eurocentrism permeating psychoanalysis resulted in a pathologizing of Buddhism, then the Orientocentrism I discussed in Chapter 2 has fostered an idealization of Buddhism. It has generated a tendency to romanticize "the East" and denigrate "the West"; to view the former as the apex of civilization, somehow transparently true and devoid of self-deception and self-blindness. Orientocentrism precludes thoughtfully examining Buddhism. It is dangerous to treat Buddhism (or any human creation) as if it were exempt from critical scrutiny and somehow free from self-interest, illusion, and corruption. Orientocentrism also insures that what psychoanalysis might offer Buddhism is neglected.

One of the unconscious functions of viewing Buddhism uncritically and ignoring what psychoanalysis might offer it, is that it allows the partiality and flaws in Buddhism to remain unscrutinized. That Buddhism has pockets of unconsciousness is suggested by such things as the residues of pathology found in enlightened meditators (cf. Brown & Engler, 1986), the plethora of scandals in spiritual communities, and the nature of unconsciousness. One reason for the resistance to psychoanalysis is that the idea of the unconscious, as Phillips (1994) aptly notes, makes a mockery of the belief in self-mastery (p. 117). After Freud, the notion of transparent self-awareness and unfettered liberation seems psychologically and morally naive. Psychoanalysis teaches us that actions and motivations that spiritual disciplines may only view from a conscious viewpoint may be riddled with more complex unconscious intentions and effects. For example, a sense of nonentitlement, self-abasement, and self-sabotage can masquerade, as it sometimes did for Albert, as spiritual nonattachment.

While exposure to Buddhist practice gave Albert the experience of dereified, decommodified, and non-self-centric subjectivity, psychoanalysis revealed some of the unconscious limitations of Buddhism.

Psychoanalytic perspectives on subjectivity helped me realize the way that Buddhism unconsciously contributed to the self-nullification of Albert, what he termed "Albertlessness." Analytic investigation revealed that Albert's attraction to the no-self doctrine of Buddhism had at least three possible meanings. First, the doctrine of no-self had an important emotional resonance for Albert because it captured his pervasive experience of nonexistence. It embodied the subjective sense of selflessness he felt about his self-nullified life. Second, the problem of self-assertion, being orphaned, was avoided by embracing this doctrine. If there was no *self*, then there was no self-*assertion*. If there was no self-assertion, then there was no threat of being ostracized and abandoned by the other. Third, the doctrine of no-self relieved the emotional pain of Albert not being who he could have become: being what he called "Albertfull". Many humans, at least in the West, desire to discover and actualize their "destiny," their unique, personal idiom (Bollas, 1989). Embracing the no-self doctrine of Buddhism anesthetized Albert's excruciatingly painful sense of having missed out on his potential.

Psychoanalysis not only clarified some of the possible origins and meanings of Albert's attraction to the Buddhist doctrine of no-self, it illuminated some of the unconscious consequences. Buddhist meditation can mollify egocentricity and decrease narcissism, even as it can also foster an *evasion* of subjectivity, which has self-betraying and self-endangering consequences. With its microscopic attention to consciousness, Buddhism helped Albert experience the texture of his inner life with greater sensitivity even as its self-negating view of subjectivity eclipsed the baby of his subjectivity in the act of throwing out the bathwater of his egocentricity.

Selves, like civilizations, are built on the twin pillars of what they exclude as well as what they include. What Kristeva (1982) terms the "abject," that which we find loathsome and other, may be a central part of what is excluded. From its place of "banishment" it exerts an uncanny significance and force. The abject is "what I permanently thrust aside in order to live.... Not me. Not that" (p. 2). Culture and individual identity rest upon it. It is "something rejected from which one does not part" (p. 4). "The place where I am not and which permits me to be" (p. 3).

"[A]bjection," according to Kristeva (1982), "is the other facet of religious, moral, and ideological codes on which rest the sleep of individuals and the breathing spells of societies. Such codes are abjection's purification and repression" (p. 209). Buddhism's telling ambivalence about emotional life may partake of the "abjection of the self" (p. 5). On the one hand, the meditative method counsels acceptance of whatever one is experiencing. On the other hand, the purpose of meditation, according to many Buddhist texts, is to "purify" the mind of "defile-ments" such as greed, hatred, and delusion. Does the tension in Bud-dhism regarding emotions, and the emphasis on cleansing impurities, signal that the self itself, in Buddhism, is viewed as abject?

Buddhism's attempted circumvention of human subjectivity does not eliminate its shaping power. A sense of nonattachment to subjective life does not mean that one will not be deeply shaped and delimited by it. The fact that sociopaths may be "detached" suggests that detachment is not freedom. One can disidentify from a troubling character trait without necessarily being free of its pernicious hold. A spiritual teacher or guru can deeply feel that the self is an illusion and that they are thus not the "owner" of any accomplishments and yet still unconsciously bask in and encourage the self-betraying submissive adulation of his or her students.

By attending to the evasion of Albert's subjectivity fostered by Buddhism, we were able to ascertain where he had unwittingly become its prisoner by perpetuating certain self-betraying modes of self-care, such as subverting and obscuring his voice, as well as reenacting restric-tive relational configurations of childhood, such as burying his discon-tent and engaging in compliant and depriving relationships that left him feeling alone and neglected.

By cultivating heightened attentiveness, detachment, and equa-nimity, meditation practice helped Albert more easily know what he was feeling, which reduced his automatic tendency to accommodate to the needs and wishes of others. It also anesthetized his emotional pain and deprivation, which stifled his affective discontent. Momentarily decentering from troubling affect inadvertently inhibited Albert from feeling the need to address some of the imprisoning aspects of his existence, which perpetuated his passivity and thereby hindered him from challenging and more fully extricating himself from the captivity of the normotic prison. We came to understand that while he had in-creasingly found the normotic mode of being of his parents, which he called "the House," vacuous and unsustaining, he had not yet built an alternative universe. Detaching from his emotional pain compromised his motivation for working on this dimension of his life.

Albert could not heal the fault line in his personality without a therapeutic relationship in which fears of individuation and self-depriving ways of caring for self and organizing relationships could be reenacted and understood and a self that reflected his values and ideals could be cultivated and developed. By systematically analyzing transference phenomena and relational reenactments, psychoanalysis can illuminate ways of being that may either go unnoticed or be submerged in Buddhism, such as Albert's idealization of his teachers and his concomitant self-submissiveness (cf. Tart & Deikman, 1991). In Buddhism this dynamic may remain unexamined, as it did with Albert, and the student's self-devaluation and deferentiality may never get resolved and may play itself out in various other relationships. This seemed to be the case with Albert in many of his relationships with both men and women.

As the crucible for the reemergence of archaic transferences, the psychoanalytic relationship can aid in the process of aborted development being recognized, reinstated, and worked through. Because Albert had only an inchoate sense of self, the Buddhist focus on deconstructing the self left him directionless even as it fostered greater awareness and tolerance of and nonattachment to thoughts, feelings, fantasies, and so forth. Since it omits the crucial task of self-construction, Buddhism's model of working with self-experience was a necessary but incomplete way of healing the fault line in his personality. He needed self-creation and self-amplification as well as self-deconstruction. Since his life was haunted by absence, emptiness, and virtuality, not misplaced desires and attachments, he needed to build a new life based on his own relational and avocational interests and ideals, not simply detach from a bad one, one based on attachments to illusory notions of self and reality. Working through a self-void and building a meaningful life is very different from letting go of illusory conceptions of self. He thus needed psychoanalysis as well as meditation in order to work through his directionlessness and build an alternative life to the normotic world.

Self-construction necessitates a somewhat different and more active role for both the analyst and the analysand than that experienced in Buddhist practice. I shall focus on four aspects of the process of facilitating authenticity that were involved in my work with Albert: (1) the importance of the analyst providing a "self-delineating selfobject function" (Stolorow, Atwood, & Brandchaft, 1992; Stolorow & Atwood, 1992), by which I mean aiding the analysand in articulating and validating his or her subjective experience; (2) the therapeutic exploration of Albert's experience of my inevitable impingements; (3) sustained investigation of his subtle affective discontent, which may signal the first faint intimations of authenticity; and (4) attention to my experience of therapeutic contentment which signified Albert's unconscious compliance in the treatment (Rubin, 1995).

The sense of self of certain, if not all, patients struggling with inauthenticity, like Albert, is tenuously held. They have scant belief in the validity of their perceptions or judgments about themselves or others. They may doubt or dismiss their own emotional reality. Their experience of emotional deprivation, for example, may be denied or minimized.

A person's sense of and confidence in the reality and substantiality of their subjective experience develops as a result of parental attunement to their subjective world, including their positive and negative affect states. There was an absence of this is Albert's life. In their designation of the "self-delineating selfobject function," Stolorow and Atwood (1992) stress the

> developmental importance of a selfobject function contributing to the articulation and validation of a child's unfolding world of personal experience ... the self-delineating selfobject function, [which] ... may be pictured along a developmental continuum, from early sensorimotor forms of validation occurring in the preverbal transactions between infant and caregiver, to later processes of validation that take place increasingly through symbolic communication and involve the child's awareness of others as separate centers of subjectivity. (1992, p. 27)

Knowledge of and belief in one's own subjective reality is deeply compromised, as we saw in Albert's case, when such parental attunement is absent. Albert struggled, for example, with self-mistrust and was prone to take on the viewpoints of others. He subordinated, to the extent of negating, his own sense of self and the world to his parents in order to maintain a sense of connectedness to them. This was a breeding ground for compliance and inauthenticity.

For an analysand who experienced absent or faulty parental responsiveness to their affective states, such as Albert, including parental denial or defensiveness about their affective malattunement, traumatic repetition of parental misattunement with the analyst is an everpresent danger. Albert assumed, for example, that sharing his inner world would lead to either enraging usurpation or hostile attacks. It is in this context that I understood his fears with me and his compensatory attempts to "speak my language."

An attuned and nonimpinging environment is essential in working with such issues. Impingements foster compliance and false self proclivities. Nonimpingement safeguards the analysand against a compliant, being-for the analyst.

Unless an analyst possessed total self-consciousness and no countertransference, which is impossible, therapeutic influence and impingement are inevitable. Analyzing their impact is thus essential. Otherwise, the perpetuation of compliance and inauthenticity goes underground. The interesting therapeutic question is not how to eliminate misattunement and impingements, which is a quixotic enterprise, but how to understand the analysand's experience of the impact of the misattunement or impingement. What is thus needed is the analyst's self-delineating self–object function, their ongoing clarification and validation of the analysand's subjective experience, including their reaction to the analyst's inevitable impingements and misattunements.

The analyst's attunement to the analysand's experience from within the latter's own vantage point is crucial to the process of clarifying and further consolidating the analysand's inchoate subjective reality. This may well include the analysand's experience of the analyst's impingement, affective discontent about the analyst or the analysis, and the analysand's vulnerability and ambivalence about experiencing and articulating their tenuously held affective reality because of fear of traumatic affective misattunement.

Once this sort of analytic environment is established, and when necessary reestablished, it is somewhat less dangerous for the analysand to experience and articulate a range of affect. This may then lead to the emergence of formally disavowed or warded off affective states.

The analysand's subtle affective discontent, for example, Albert's irritation and muted disappointment in me for not affirming his ideas, can represent the first faint expression of authenticity, which will probably be experienced by him or her as unfamiliar, trivial, strange, or silly. The analysand may (1) not believe in, (2) dismiss, or (3) devalue moments of authenticity just as important caregivers failed to value and encourage the expression of and belief in those aspects of the analysand's subjective world at earlier stages of his or her development. Albert experienced all three at different times in our work together. My interest in these intimations of authenticity aided in their further expression and clarification. This in turn, eventually led to Albert's strengthened recognition of and belief in their existence and validity.

Attunement to the analyst's experience, particularly his or her sense of therapeutic contentment, as well as the analysand's discontentment, can sometimes also aid in facilitating the emergence of the analysand's authenticity. For example, my sense of excessive efficacy and ease at various points in the treatment signaled not that the analysis was proceeding in an unequivocally successful manner, but that it was permeated by Albert's unconscious compliance with me. Exploring moments in which Albert tried to "speak my language" because of the emotional dangers that he anticipated eventually opened up the pattern of self-submissiveness inculcated with his parents that permeated his relationships, professional life, and the treatment. The feeling that I had that I was in tune with Albert hid the way in which he was sometimes accommodating to me. My sense of therapeutic success was periodically an artifact of his conformity rather than the correctness of my interpretations. Attention to this dimension of my experience aided me in reexamining subtle aspects of compliance in the therapeutic interaction, which encouraged transference analysis of this formerly neglected area. Albert described our work in an association during this period of treatment as "a crane pulling the buried coffin out of the earth."

We were then able to begin to address other issues such as the obstacles he experienced and established to intimacy and professional fulfillment, which Buddhism neglects. Spiritual seekers are then often at a loss in these areas.

As Albert was more able to be authentic with me, his personal and professional relationships also began to change. He formed more reciprocal social and intellectual partnerships and his friendships and his relationship with his wife were more fulfilling.

His professional work was also enriched. As he valued his own needs more, he became more comfortable with self-assertion, and his writing, like his life, became more "Albertfull." Tired of keeping his potential under wraps and seducing others into diminished views of his academic work, he was able to put his best intellectual foot forward, which resulted in his receiving acknowledgment for his academic talents by those he valued in his field.

Non-self-centered subjectivity and self-centered subjectivity are states that anyone can experience. Often they are in tension within an individual. It is affectively difficult to keep this tension alive. Collapsing this core dialectic in human psychology by valorizing either experience of self affords psychic relief even as it impoverishes our understanding of human psychology (Charles Spezzano, personal communication).

As I hope my work with Albert has suggested, psychoanalysis and Buddhism were *both* necessary to keep alive his tenuous hope for a life that might feel more authentic and alive. Psychoanalysis' illumination of the historical sources of his difficulties in living, such as the effects of unempathic parenting on self-development (cf. Kohut, 1971, 1977, 1984) and unconscious commitments to old and restrictive interpersonal ways of being (e.g., Fairbairn, 1952), helped Albert loosen the grip of his limiting past. Buddhism's elucidation of the sources of suffering in the present, such as his attachment to restrictive conceptions of self, also aided Albert in breaking free from an imprisoning life. Through his experience in analysis and meditation, Albert was able to relate to affect differently, become less self-critical, improve his self-care, learn new skills and pursue new avocations, and relate more authentically and sensitively to others.

Buddhism's view of non-self-centric subjectivity tempers psycho-analysis' egocentricity and widens its moral purview. Psychoanalysis illuminates some of the deleterious psychic consequences of Bud-dhism's evading of subjectivity. Psychoanalysis teaches Buddhism that there is a hidden and pernicious cost to absolutizing its view of the fictionality of the self, especially a "return of the repressed" that may generate pain in its wake.

Albert's experience with analysis and Buddhism illustrates my the-oretical claim in Chapter 3 that psychoanalytic and Buddhist concep-tions of subjectivity would be enriched if the understandings obtained from their different ways of investigating subjectivity were integrated into a more encompassing and inclusive framework. A more encom-passing model of subjectivity would illustrate a more complete range of the multiplicity that is the self than either psychoanalytic or Buddhist conceptions pursued alone. Adopting such a complementary bifocal framework about self and nonself enabled Albert to recognize that states of self-centeredness and unselfconsciousness were both part of his at-tempts to live a full and meaningful life. The former was necessary to help him fixate the self and view it as a concrete, substantial entity. This helped him reflect on his life and conduct, delineate what he felt and valued, assess situations, formulate plans and goals and choose among potential courses of action. It helped him ultimately find an alternative to the suffocating normotic world of his parents. Sensitivity to states of non-self-consciousness enabled him to live less self-centeredly, more flu-idly and gracefully. The view of the self as a process facilitated his appreciation of art, his capacity to listen to his students, play music, and experience love.[1]

REFERENCES

Bollas, C. (1987). *The shadow of the object: Psychoanalysis of the unknown thought.* New York: Columbia University Press.
Bollas, C. (1989). *Forces of destiny: Psychoanalysis and human idiom.* London: Free Association Books.
Brown, D., & Engler, J. (1986). The stages of mindfulness meditation: A validation study. In K. Wilber, J. Engler, & D. Brown (Eds.), *Transformations of consciousness: Conven-tional and contemplative perspectives on human development* (pp. 17– 51). Boston: Shambhala.

[1]Dialogues with George Atwood, Jerry Gargiulo, James Jones, Marilyn Saur, Charles Spez-zano, Deborah Tanzer, and Mary Traina enriched this chapter. I am grateful to Jack Korn-field for his integrative perspective on the "factors of enlightenment," which was enor-mously helpful as a template for integrating psychological and spiritual practices.

Buccino, D. (1993). The commodification of the object in object relations theory. *Psychoanalytic Review, 80*(1), 123–134.

Butler, K. (1990, May/June). Encountering the shadow in Buddhist America. *Common Boundary*, pp. 14–22.

Butterfield, S. (1994). *The double mirror: A skeptical journey into Buddhist Tantra.* Berkeley, CA: North Atlantic Books.

Engler, J. (1986). Therapeutic aims in psychotherapy and meditation: Developmental stages in the representation of self. In K. Wilber, J. Engler, & D. Brown (Eds.), *Transformations of consciousness: Conventional and contemplative perspectives on human development* (pp. 17–51). Boston: Shambhala, 1986.

Epstein, M. (1995). *Thoughts without a thinker: Psychotherapy from a Buddhist perspective.* New York: Basic Books.

Fairbairn, R.D. (1952). *Psychoanalytic studies of the personality.* London: Routledge and Kegan Paul.

Fauteux, K. (1987). Seeking Enlightenment in the East: Self-fulfillment or regressive longing? *Journal of the American Academy of Psycho-Analysis, 15,* 223–246.

Finn, M. (1992). Transitional space and Tibetan Buddhism: The object relations of meditation. In M. Finn & J. Gartner (Eds.), *Object relations theory and religion: Clinical applications* (pp. 109–118). Westport, CT: Praeger.

Freud, S. (1917). A difficulty on the path of psycho-analysis. In J. Strachey (Ed. & Trans.), *The standard edition of the complete psychological works of Sigmund Freud,* Vol. 17 (pp. 135–144). London: Hogarth Press.

Freud, S. (1933). New introductory lectures. In J. Strachey (Ed. & Trans.), *The standard edition of the complete psychological works of Sigmund Freud,* Vol. 22 (pp. 5–182). London: Hogarth Press.

Gaylin, W. (1988). Love and the limits of individualism. In W. Gaylin & E. Person (Eds.), *Passionate attachments: Thinking about love* (pp. 41–62). New York: Free Press.

Gedo, J. (1986). *Conceptual issues in psychoanalysis.* Hillsdale, NJ: Analytic Press.

Gill, M. (1982). *The analysis of transference,* Vol. 1. New York: International Universities Press.

Goldstein, J. (1976). *The experience of insight: A natural unfolding.* Santa Cruz, CA: Unity Press.

Goleman, D. (1977). *The varieties of the meditative experience.* New York: Dutton.

Hartmann, H. (1960). *Psychoanalysis and moral values.* New York: International Universities Press.

Heimann, P. (1950). On counter-transference. *International Journal of Psycho-Analysis, 31,* 81–84.

Heimann, P. (1960). Counter-transference. *British Journal of Medical Psychology, 33,* 9–15.

Kipling, R. (1942). *A Kipling pageant.* Garden City, NY: Halcyon House.

Kohut, H. (1971). *The analysis of the self.* New York: International Universities Press.

Kohut, H. (1977). *The restoration of the self.* New York: International Universities Press.

Kohut, H. (1984). *How does analysis cure?* Chicago: University of Chicago Press.

Kohut, H., & Wolf, E. (1978). The disorders of the self and their treatment. *International Journal of Psycho-Analysis, 59,* 413–425.

Kornfield, J. (1977). Living Buddhist masters. Santa Cruz, CA: Unity Press.

Kornfield, J. (1979). Psychological adjustment is not liberation. *Zero: Contemporary Buddhist Life and Thought, 2,* 72–87.

Kornfield, J. (1993a). *A path with heart: A guide through the perils and promises of spiritual life.* New York: Bantam Books.

Kornfield, J. (1993b). The seven factors of Enlightenment. In R. Walsh & F. Vaughan (Eds.),

Paths beyond ego: The transpersonal vision (pp. 56–59). Los Angeles, CA: Tarcher/ Perigee.

Kramer, J., & Alstad, D. (1993). *The guru papers: Masks of authoritarian power*. Berkeley, CA: North Atlantic Press.

Kristeva, J. (1982). *Powers of horror: An essay on abjection*. New York: Columbia University Press.

Levinson, E. (1983). *The ambiguity of change*. New York: Basic Books.

Panel. (1984). Value judgments in psychoanalytic theory and practice (S. Lytton, reporter). *Journal of the American Psychoanalytic Association, 32*, 147–156.

Phillips, A. (1994). *On flirtation: Psychoanalytic essays on the uncommitted life*. Cambridge, MA: Harvard University Press.

Roland, A. (1988). *In search of self in India and Japan: Towards a cross-cultural psychology*. Princeton, NJ: Princeton University Press.

Rubin, J.B. (Submitted). *The blindness of the seeing I: Perils and possibilities in psychoanalysis*. New York: NYU Press.

Rushdie, S. (1994). *East, West*. New York: Pantheon Books.

Stolorow, R., & Atwood, G. (1992). *Contexts of being: The intersubjective foundations of psychological life*. Hillsdale, NJ: The Analytic Press.

Stolorow, R., Atwood, G., & Brandchaft, B. (1992). Three realms of the unconscious and their therapeutic transformation. *Psychoanalytic Review, 79*, 25–30.

Stricker, G., & Gold, J. (1993). Psychotherapy integration with character disorders. In G. Stricker & J. Gold (Eds.), *Comprehensive textbook of psychotherapy integration* (pp. 323–336). New York: Plenum Press.

Sullivan, H.S. (1953). *The interpersonal theory of psychiatry*. New York: Norton.

Tart, C., & Deikman, A. (1991). Mindfulness, spiritual seeking and psychotherapy. *Journal of Transpersonal Psychotherapy, 23*(1), 29–52.

Thera, S. (1962). *The way of mindfulness: The Satipatthana Sutra and commentary*. Kandy, Sri Lanka: Buddhist Publication Society.

Varela, F. (1984). The creative circle: Sketches on the natural history of circularity. In P. Watzlawick (Ed.), *The invented reality* (pp. 309–323). New York: Norton.

Wachtel, P. (1977). Psychoanalysis and behavior therapy. New York: Basic Books.

10

Toward a Contemplative Psychoanalysis

Psychoanalysis and Buddhism are stories about and strategies for addressing human life. Treating Buddhism and psychoanalysis as narratives rather than as sacred tradition, e.g., sources of absolute wisdom that provide a blueprint for how to live in the present, may shift the way we think about tradition in general and each tradition in particular. Instead of viewing either of them as "received truths," universally valid for all times and places, we might conceive of them as human creations arising in particular historical and sociocultural contexts whose value resides in how well they help people in the present age live with greater awareness, tolerance, and care.

Tradition has two meanings: it means to pass on and it means to betray. Tradition can be enslaving as well as enabling. It may give one an identity and an orientation in the world, even as it delimits one's horizon of vision and stifles one's development. It is inhibiting because it assimilates the present into the past and predisposes us to look toward the past to solve dilemmas in the present. "Tradition is important," as Joel Kramer (1977) aptly notes, "just as history is important—not as a vise to squeeze the present into, but as a steppingstone to grow from" (p. 26).

Once tradition is no longer viewed as sacred, its essential revisability becomes more crucial. Buddhism, as well as psychotherapy or psychoanalysis, needs to be open to feedback about its limits and to change, evolve, and grow so that it can respond to the living moment.

There are signs that psychoanalysis and Buddhism suffer from being static and that they are evolving. The increasing attempt to question the authority of Buddhist teachers, democratize spiritual communities, and make Buddhism more "engaged" with social issues, such as the peace movement, and the heightened emphasis on the value of countertransference and the therapeutic relationship in psychoanalytic

189

treatment illustrate the way both traditions are evolving. The neglect of spirituality and health in psychoanalysis and subjectivity, affective life, and historicity in Buddhism illustrate the need for revision in both traditions.

Let me give one example from Buddhism and psychoanalysis of the way each needs to be open to new influences in order to meet the needs of late twentieth-century people. Right livelihood (work that helps rather than hurts human life) is one of the eight aspects of the Eightfold Path to liberation in Buddhism. Tax collectors, as well as butchers and prostitutes, were traditionally exemplars of wrong livelihood in classical Buddhism. Perhaps tax collectors were destructive to human life in fifth-century BC India. Some people might argue that they are still destructive to human life in our age, but if tax collectors were abolished in our world, then social service programs that attempt to help people in need would also suffer (Joel Kramer, personal communication).

Psychoanalysis offers a critique of the isolated, autonomous, Cartesian subject. Yet it often takes the self-centered, egocentric self as the touchstone of its views of morality. In a world of increasing scarcity and selfishness, psychoanalysis might benefit from reformulating the subject or self that it works on, perhaps developing a less self-centered sense of self. What both examples illustrate, and readers could undoubtedly find other examples, is that uncritically grafting the doctrines of one tradition onto another from another age can be problematic. Because Buddhism and psychoanalysis often seem to assume the universal validity of their theories, it may be difficult for them to consider the possibility that they would benefit from updating traditional formulations such as the Eightfold Path or the self-centered self so as to be attuned to historical and cultural conditions and needs that are qualitatively different than fifth-century BC India or nineteenth-century Europe. Conditions in late twentieth-century Western life, as I have suggested in earlier chapters, may necessitate alternative theories.

The anthropological study of psychiatric disorders has yielded a multitude of conditions that are found only in one or another culture such as a syndrome known in Latin American cultures as *susto*, or "loss of the soul," and a Japanese malady known as *taijin kyofusho*. In *susto*,

---

[1]There are, of course, exceptions such as the cross-cultural work of Roland (1988). A sensitivity to cross-cultural psychological differences is also evident in the work of scholars in fields such as anthropology (e.g., Richard Shweder, 1991). Unfortunately, most psychoanalysts and Buddhists do not seem to have explored or integrated this work, sometimes labeled "cultural psychology," into their own disciplines. Its potential impact on traditions such as psychoanalysis and Buddhism thus remains to be more fully developed.

the soul of the surviving person departs with the dead person, leaving the survivor soulless. *Taijin kyofusho* involves fear of people, especially social shame, a morbid dread of embarrassing other people (cf. Goleman, 1995). If mental illness varies, at least in part, with culture, then it may be distorting to assume a cross-cultural uniformity to psychological phenomenon. The value of psychic particularity as opposed to psychic universality and homogeneity has important implications for the integration of psychotherapeutic and contemplative traditions in general and psychoanalysis and Buddhism in particular. For if there are conditions that are unique to Western, industrialized, individualistically oriented culture, such as anorexia nervosa or the depleted, alienated, oversaturated self, then it might be simplistic to believe that theories and methods developed in other cultures and historical ages to address a different and unique set of historical, sociological, and psychological exigencies have all the answers for our unique context.

Contemporary America, like fifth-century BC India, is in a time of ferment and turmoil. There is a pervasive sense of deprivation, disconnection, alienation, and nihilism. Skepticism about politics and religion and the breakdown of civil society impoverish our emotional lives and generate emptiness and suffering.

How we respond to our crisis is crucial, for where we are as a species is a direct outgrowth of who we are. It is feedback that old and familiar ways of being, including aspiritual psychologies and apsychological, morally simplistic spiritualities, are not working.

We are a divided species. On the one hand, we explore the outer limits of the solar system and the vast expanse of inner space. On the other hand, our political leaders pursue social policies that demonstrate little regard for the welfare of the people they are purported to represent and the citizenship demonstrates an apathy, myopia, and deferentiality that is frightening. Seeking immediate pleasures, self-centered, loath to assume responsibility, escaping from pain (through endless soporifics such as acquiring and consuming, alcohol, drugs, spectator sports, soap operas, and exhibitionist talk shows), and oblivious to the perils of its myopic mode of being, late twentieth-century America as a culture seems as if it were in its adolescence.

To respond to the challenges of our times, we need something new. Our health and our survival demand that we meet the problems that confront us as adults. Adulthood involves, among other things, taking what we inherit, psychologically and morally, including psychoanalysis and Buddhism, and making it our own. Adults demonstrate the capacity to be open to and transformed by feedback so that they can respond to the unique challenges that confront them freshly and creatively.

Certain aspects of psychoanalytic and Buddhist theory or doctrine are sediments of tradition that compromise our capacity to respond to the unique nature of the living moment. Instead of responding to the flux and unpredictability of life in the present moment, loyalty to tradition, to the past, to what we already know, takes precedence. This delimits our responsiveness.

Challenging taken-for-granted modes of thought and being, the methods of both psychoanalysis and Buddhism offer tools to respond to the present in an unfettered and innovative way. The self-reflexive dialogue of analyst and analysand and the special way of speaking and listening fostered by the patient's free associations and the analyst's evenly hovering attention are, as I have suggested in Chapter 1, an essential facet of the psychoanalytic method. Meditation, the careful, attentiveness to all aspects of one's thoughts, feelings, fantasies, and conduct, what a Vietnamese Zen teacher termed "mindful living" (Hanh, 1989), is the indispensable core of Buddhist method.

Let me return to the theme of the beginning of this chapter and reflect briefly on the stories psychoanalysis and Buddhism tell about the nature of human life. Stories are made, as historian and cultural theorist and critic Hayden White (1973) notes, "by including some events and excluding others, by stressing some and subordinating others" (p. 6, note 5). "Emplotment" is what White terms this process of exclusion, emphasis, and subordination in the interest of creating a particular kind of story. Literary theorist Northrop Frye (1957) has identified four archetypal genres or types of plot structures-tragedy, irony, romance, and comedy. (Since there may be other modes such as the epic, and a story may contain a combination of modes, I am viewing Frye's cartography as suggestive rather than comprehensive.) Each genre offers a conception of the world that is particular and partial, highlighting and omitting certain facets of the world.

With its recognition of the inescapable mysteries, dilemmas, and afflictions pervading human existence, psychoanalysis is underwritten by a "tragic" worldview (Schafer, 1976). Tragic does not necessarily imply, contrary to popular usage, "unhappy or disastrous outcomes" (Schafer, 1976, p. 47), but rather a steadfast recognition that time is irreversible and unredeemable; humans are beings moving toward death, not rebirth; choices entail conflict and compromise; and self-division and sacrifice are inevitable. Religious consolations are quixotic in the tragic vision.

Steadfastly acknowledging the ubiquity of suffering, Buddhism appears to be emplotted in a tragic mode. Given the emphasis on the pervasiveness of suffering in Buddhism, it thus seems odd—if not terribly distorting—to claim that Buddhism is emplotted in a romantic narrative. Buddhism's *diagnosis* of the human condition is tragic. But its *prognosis*—as I attempted to demonstrate in Chapter 4, "The Emperor of Enlightenment May Have No Clothes"—is romantic. With its belief in the possibility of self-mastery, transcendence, and unqualified fulfillment—the possibility of transcending what Shakespeare termed "the thousand shocks that flesh is heir to"—Buddhism is a romantic narrative about human existence. Romance refers not to romantic involvement, infatuation with another, or idealized love, but rather to a view of the world that assumes that personal and familial conditioning can be transcended and that ultimate meaning on a grand design can be achieved. In the romantic vision, as Frye (1957) notes, life is viewed as a quest involving the hero's or heroine's "transcendence of the world of experience, his [or her] victory over it, and his [or her] final liberation from it" (p. 8).

There are no neutral or objective grounds to evaluate the ultimate validity for either set of stories. There is no way either story could be "refuted" or "disconfirmed" by appeal to the facts of history or current experience, because how we interpret and evaluate history and human experience is deeply conditioned and irreducibly shaped by the worldview we already hold. Someone with a tragic worldview, for example, would see the world perilously perched on the brink of destruction. Someone with a more romantic perspective might maintain that our collective malaise is a sign of the bankruptcy of antiquated modes of being and the harbinger of transformation and enlightenment. The stories psychoanalysis and Buddhism tell thus represent alternative and incommensurable views of human existence. The grounds for choosing either story are thereby *pragmatic* and *aesthetic*, not epistemological. One important criterion of which story we choose is what we are looking for—what kind of world we wish to inhabit and what kind of life we want to live.

We live in difficult—even dangerous—times. We are at a cusp point: Our world flirts with planetary ecocide, and our future is uncertain. The Chinese ideogram for crisis is challenge *and* opportunity. The way we respond to our planetary crisis will play an important role in determining our collective future.

The times demand both a sobering recognition of the fragility and tenuousness of our condition and a decisive response to the enormous challenges that collectively face us. Optimism is a better strategy for change than pessimism (Joel Kramer, personal communication). Pessimism often breeds paralysis, which inhibits the motivation to change. But ungrounded optimism can result in an illusory and disabling conception of reality. If skepticism without hope is enervating, then optimism without realism is quixotic.

With his notion of "pessimism of the intellect, optimism of the will," Gramsci (1971, p. 175, note 75) provides one possible way of integrating the stories psychoanalysis and Buddhism tell so that they might speak to the concerns of late twentieth-century citizens confronting meaninglessness, disconnection, self-alienation, and so forth.

Unbridled optimism can be an opiate for reformers. The absence of pessimism opens the door to illusory and self-deceptive hopes about the possibilities for transformation and thus substitutes for the painful and necessary job of confronting the pain, injustice, and moral exhaustion in our world. But pessimism, on the other hand, easily slides into paralysis and can thus hinder imagination. "Insofar as social theory declines to be romantic," it is, as Rorty (1991) notes, "inevitably retrospective, and thus biased towards conservatism" (p. 188). Castoriadis (1987) speaks of the "contingency ... poverty and ... insignificance" (p. 155) of the existing system of imaginary significations constituting society, with its social arrangements and descriptions of personal and collective possibility. These ways of conceiving reality and the social world are "nothing more than 'frozen politics' ... they serve to legitimate, and make seem inevitable, precisely the forms of social life ... from which we [North American liberals] desperately hope to break free" (Rorty, 1991, p. 189).

Imagination born of romanticism can offer alternatives to "the present system of imaginary significations" (Rorty, 1991, p. 192). Different futures are often hatched in imagination—in, for example, dreams, novels, political essays, or science fiction. Imagination can prevent us from giving, in Unger's (1986) phrase, "the last word to the historical world we inhabit" (pp. 118–119). The values, language, and social and political arrangements of the time we live in may then not have to completely delimit the horizon of how we might see, describe, and arrange the world. The imaginative reflections of a person with a romantic sensibility can fashion possibilities where before one foresaw only limitations.

With its tragic conception of the universe and its "hermeneutics of suspicion" (Ricoeur, 1970)—its challenging and demystifying of motives and meanings—psychoanalysis embodies "pessimism of the intellect." With its belief in radical, thoroughgoing liberation and its active practice of transformation, Buddhism is an exemplary instance of "optimism of the will." We need an awareness of both the tragic realities and the romantic possibilities in our world. The trick is to neither adopt romantic delusions that dilute the sober realities we confront nor fall into a pessimistic abyss about the dearth of possibilities. Buddhism needs psychoanalysis' qualification of its romantic aspirations. Psychoanalysis' strategies for working with self-unconsciousness tempers Buddhism's romanticism. Buddhism's method of cultivating exquisite moment-to-moment attentiveness, quieting and clarifying the inner pandemonium, and fostering nonattachment (not detachment) challenge psychoanalysis' pessimistic view of human possibilities. There is more to life, suggests Buddhism, than the common human unhappiness Freud felt was the apex of human health. If Buddhism minimizes the extent of human self-blindness, than psychoanalysis does not acknowledge the possibilities for self-transformation.

A judicious combination of what Buddhism posits about the possibilities for self-transformation and liberation and what psychoanalysis asserts about the complexity of self-investigation and the difficulties of self-transformation can aid us in avoiding despair and ungrounded romanticism. Integrating the stories psychoanalysis and Buddhism tell about human life might provide late twentieth-century citizens with a wider range of tools for coping with the unique historical exigencies of our world.

In the conclusion of this chapter, I shall attempt to paint a picture of a contemplative psychoanalysis, a therapeutics that draws on the valuable facets of psychoanalysis and Buddhism.

TOWARD A CONTEMPLATIVE PSYCHOANALYSIS

A contemplative psychoanalysis of the future might explore both what psychoanalysis might contribute to Buddhism and other spiritual traditions and what Buddhism and other contemplative traditions might teach psychoanalysis. A therapeutics that judiciously draws on spiritual practices of training attentiveness and fostering nonattachment or a nonnarcissistic relation to our own thoughts, beliefs, and lives such as Buddhism offers (cf. Rubin, 1985, 1995) and psychoanalytic modes of investigating self-experience and interpersonal relatedness might enable us to distinguish between and reconcile the possibilities and pathologies of self-centered and non-self-centered subjectivity. It might thus deabsolutize the self-centered self and deprovincialize the non-self-centered self.

In terms of the former, when we are less attached to an overly self-centered sense of self, we may feel released from the constraining grip of a self-alienating egoism, we may experience a greater capacity for empathy and tolerance, and we may feel a heightened sense of connection with and compassion for others. In such a self-forgetful state as occurs, for example, when one is deeply immersed in nature, creating art, or making love, one experiences a heightened sense of living. Greater intimacy may flow from such an experience of subjectivity, since in friendship and love we need to loosen and sometimes become free of our normally more delimited, self-referential sense of separateness from others and the world.

Non-self-centered subjectivity might also enrich analytic treatment in several ways. When in a less bounded, non-self-preoccupied, non-self-centric state, the analyst can more easily engage in transient identifications with subtle affect experiences, access somatic knowledge,[2] and decenter from habitual and limiting psychological conditioning. In this state there is sometimes "An easy commerce of the old and the new" (Eliot, 1963, pp. 207–208), resulting in what the Indian philosopher Krishnamurti (1969) termed a "freedom from the known," in which we are able to spring from familiar patterns of thinking and relating and experience newness. It is illustrated by the analyst who makes a creative interpretation or intervention, saying something she was not consciously aware that she knew or relating in an unfamiliar yet creative and facilitative manner, instead of reiterating something unoriginal that she already knows. These states are likely to promote in the analyst increased capacity to listen creatively and respond to the exigencies of the clinical moment in a less fettered manner.

Analysands who experience non-self-centered subjectivity may have less difficulty alternating between what Modell (1989) terms the "multiple realities" of the analytic setting and process, such as past and present, transference as "real" and illusory, that are essential to change. The analysand's capacity for psychic fluidity would probably also facilitate the capacity to free associate.

The non-self-centered subjectivity characterizing Buddhism might become deprovincialized when Buddhism welcomes the questions analysts might have about such things as the potential pathologies of nonattachment, self-evasion, spiritual idealization, and submissiveness. Self-betraying facets of Buddhism including the abjectification of self and emotions, the self-compromising facets of the hidden authoritarianism of the teacher–student relationship, and the potential coun-

---

[2]Somatic knowledge is illustrated when we "remember" the PIN number or gym lock combination we have momentarily forgotten. Remembering occurs not when we try to think about the number, which only seems to generate more frustration and temporary amnesia, but when we "let our fingers do the walking" and thinking.

tertransference of Buddhist teachers might then become the focus of analytic and meditative inquiry.

The moral iniquity and vacuity of current civic and public life and the absence of thoughtful and substantive dialogue about how education, media, families, and public and religious institutions might celebrate diversity, strengthen bonds between peoples, and nourish life-affirming modes of being and relating cry out for progressive responses. In facilitating our recognition of the reality of moral complexity and human agency, while not eclipsing the capacity for non-self-centered subjectivity, a contemplative psychoanalysis might be more ethically responsive to the moral challenges humans now face than either psychoanalysis or a spiritual discipline such as Buddhism pursued alone. Out of a contemplative psychoanalysis we could fashion a permeable but bounded postindividualistic, "species self" (cf. Lifton, 1993) for whom both individuation and non-self-centricity are seen as two interpenetrating facets of what it means to be a human being. A new and expanded sense of I-ness fostering links between people, rather than a privatist, individualist ethos, might be one such beacon in our collective darkness. A spiritually attuned contemplative psychoanalysis might see the needs of fellow citizens and the polis as crucial organizing principles for the conduct of human lives.

The sphere of empathy of a species self would reach beyond oneself and those significant others within one's circle of concern to include fellow citizens of the world and the dispossessed. Seeing oneself in a more expansive and interconnected yet differentiated way, as a self-in-community instead of a selfless spiritual self or an isolated, imperial psychological self, might foster both connectedness to the polis and self-enrichment and thus decrease alienation and anomie.

The beliefs and practices of a contemplative psychoanalysis would themselves be open to examination and, where necessary, transformation. It would thus be an open system, an apparent contradiction. Systems are ordinarily structures that are closed and have a slow rate of change. An open system would value feedback from those people it serves and those who theoretically challenge it, and be open to self-correction and transformation.[3]

There is a growing and disturbing disparity in the 1990s between the haves and the have-nots. By fostering a sense of connection with and tolerance for the Other, spirituality may be a crucial aspect of our collective mental health, as long as it is open to the questions of psycho-

---

[3]I owe this conception to the work of Bruce Lee and Dan Inosanto, who created an open system of martial arts called Jeet Kune Do.

analysis. When the spiritual is no longer delegitimized, as it has traditionally been in psychoanalysis,[4] then there might be a "widening scope" of psychoanalysis that might encourage analytically informed examinations of issues that adversely affect the health of cultures as well as individuals: the seductions of materialism and consumerism and the psychology of greed, authoritarianism, submissiveness to fanaticism, racism, sexism, homophobia, and ageism. Questions about such things as the nature of creativity, health, and intimacy, engaged citizenship, and the process of fostering psychological decolonization might then be seen as more germane to our field. A psychoanalysis that was not only self-centered, that was informed by contemplative perspectives, might contribute to a civilization with less discontent.[5]

REFERENCES

Castoriadis, C. (1987). *The imaginary institution of society.* Cambridge, MA: MIT Press.

Eliot, T.S. (1963). Little Gidding. In T.S. Eliot (Ed.), *Collected poems (1909–1962)* (pp. 207–208). New York: Harcourt, Brace & World.

Fromm, E., Suzuki, D.T., & DeMartino, R. (Eds.) (1970). *Zen Buddhism and psychoanalysis.* New York: Harper & Row.

Frye, N. (1957). *Anatomy of criticism: Four essays.* Princeton, NJ: Princeton University Press.

Goleman, D. (1995, December 5). Making room on the couch for culture. *New York Times,* pp. C1–C3.

Gramsci, A. (1971). *Selection from the prison notebooks* (Q. Hoare & G. Smith, Eds. & Trans.). New York: International Universities Press.

Hanh, T. N. (1989, November/December). Seeding the unconscious: New views on Buddhism and psychotherapy. *Common Boundary,* pp. 14–21.

Horney, K. (1945). *Our inner conflicts.* New York: Norton.

Horney, K. (1987). *Final lectures.* New York: Norton.

Jung, C.G. (1958). *Psychology and religion: West and East.* Princeton, NJ: Princeton University Press.

Kelman, H. (1960). Psychoanalytic thought and Eastern wisdom. In J. Ehrenwald (Ed.), *The history of psychotherapy* (pp. 328–333). New York: Jason Aronson.

Kramer, J. (1977). Playing the edge of mind and body: A new look at yoga. *The Yoga Journal, January,* 26–30.

Krishnamurti, J. (1969). *Freedom from the known.* New York: Harper & Row.

[4]There are, as I suggested in Chapter 2, notable exceptions such as Horney (1945, 1987), Jung (1958), Kelman (1960), Fromm, Suzuki, and DeMartino (1970), Roland (1988), and Rubin (1985, 1991, 1993). But the cross-fertilization that might mutually enrich both disciplines tends to be lacking even in approaches that avoid Eurocentrism and Orientocentrism.

[5]This chapter benefitted from discussions with Bettina Edelstein, Jim Jones, and Eugene Murphy.

Lifton, R. (1993). *The protean self: Human resilience in an age of fragmentation*. New York: Basic Books.

Modell, A. (1989). The psychoanalytic setting as a container of multiple levels of reality: A perspective on the theory of psychoanalytic treatment. *Psychoanalytic Inquiry, 9*(1), 67–87.

Ricoeur, P. (1970). *Freud and philosophy: An essay on interpretation.* (D. Savage, Trans.). New Haven, CT: Yale University Press.

Roland, A. (1988). *In search of self in India and Japan: Towards a cross-cultural psychology*. Princeton, NJ: Princeton University Press.

Rorty, R. (1991). *Essays on Heidegger and others*. New York: Cambridge University Press.

Rubin, J.B. (1985). Meditation and psychoanalytic listening. *Psychoanalytic Review, 72*(4), 599–612.

Rubin, J.B. (1991). The clinical integration of Buddhist meditation and psychoanalysis. *Journal of Integrative and Eclectic Psychotherapy, 10*(2), 173–181.

Rubin, J.B. (1993). Psychoanalysis and Buddhism: Toward an integration. In G. Stricker & J. Gold (Eds.), *Comprehensive textbook of psychotherapy integration* (pp. 249–266). New York: Plenum Press.

Rubin, J.B. (Submitted). *The blindness of the seeing I: Perils and possibilities in psychoanalysis*. New York: NYU Press.

Schafer, R. (1976). *A new language for psychoanalysis*. New Haven, CT: Yale University Press.

Shweder, R. (1991). *Thinking through cultures: Expeditions in cultural psychology*. Cambridge, MA: Harvard University Press.

Ungar, R. (1986). *The critical legal studies movement*. Cambridge, MA: Harvard University Press.

White, H. (1973). *Metahistory: Historical imagination in nineteenth-century Europe*. Baltimore, MD: John Hopkins University Press.

Index

About the Author

For the past 18 years, Dr. Jeffrey B. Rubin has practiced psychoanalysis and Buddhist meditation. He has taught psychoanalysis and Buddhist theory and practice at Yeshiva University and Goddard College. He is on the faculty of two psychoanalytic institutes: Postgraduate Center for Mental Health and the C. G. Jung Foundation in New York City. He has published papers on subjects ranging from the politics of psychoanalysis to meditation and psychoanalytic listening. Dr. Rubin is author of the forthcoming *The Blindness of the Seeing I: Perils and Possibilities in Psychoanalysis*. He practices psychoanalysis and psychoanalytically oriented psychotherapy in New York City and Bedford Hills, New York.